A2 Geography
UNIT 5

Edexcel

Specification **B**

Unit 5: Researching Global Futures
(Living with Hazardous Environments)

Sue Warn and David Holmes

Philip Allan Updates
Market Place
Deddington
Oxfordshire
OX15 0SE

Orders

Bookpoint Ltd, 130 Milton Park, Abingdon, Oxfordshire, OX14 4SB
tel: 01235 827720
fax: 01235 400454
e-mail: uk.orders@bookpoint.co.uk
Lines are open 9.00 a.m.–5.00 p.m., Monday to Saturday, with a 24-hour message answering service. You can also order through the Philip Allan Updates website: www.philipallan.co.uk

© Philip Allan Updates 2007

ISBN 978-1-84489-005-7

This guide has been written specifically to support students preparing for the Edexcel Specification B A2 Geography Unit 5 examination. The content has been neither approved nor endorsed by Edexcel and remains the sole responsibility of the authors.

Printed by MPG Books, Bodmin

Philip Allan Updates' policy is to use papers that are natural, renewable and recyclable products and made from wood grown in sustainable forests. The logging and manufacturing processes are expected to conform to the environmental regulations of the country of origin.

Contents

Introduction

■ ■ ■

Content Guidance

■ ■ ■

Questions and Answers

Introduction

About this guide

The purpose of this guide is to help you prepare for Unit 5: Researching Global Futures, Option 5.2: Living with Hazardous Environments. The guide is divided into three sections.

This **Introduction** outlines the structure of the specification, the assessment procedure and the use of pre-released 'generalisations'.

The **Content Guidance** section first covers how to carry out effective research and then sets out the bare essentials of the specification for this option. A series of diagrams is used to help your understanding; many are simple to draw and could be used in the exam.

The **Question and Answer** section begins with guidance on how to write good quality essays about hazards. It also provides an extensive list of sample questions for each of the generalisations. For one of these, two sample answers of differing standards are provided, marked according to the generic mark scheme, together with examiner's comments.

The specification

This unit involves a combination of classroom study and student research. The specification starts with the *Foundation for Study* and is followed by four generalisations, designed to take equal amounts of study time. There is much interlinkage between the generalisations.

The specification is divided into three sections:
- **Key enquiry questions** These should be the starting point for your studies. The questions are sequenced so you can build up your understanding as you progress.
- **Guidance for students** This gives ideas on selecting case-study materials from a variety of locations (including countries at various states of development) and at a range of scales. Note that these are intended for *guidance only*. It is essential that you research up-to-date examples (e.g. in 2005, Katrina, Kashmir and the aftermath of the Boxing Day Tsunami or in 2006, the contrasting droughts of southeast England and central Australia).
- **Generalisations** These are ideas or concepts that you need to grasp. Examiners set essay titles based on these generalisations. The title of the chosen generalisation is released about 6 weeks before the exam.

Getting started

You must start with the *Foundation for Study*, which introduces a number of key definitions and concepts, It reminds you to focus on *natural* hazards and introduces the major hazard categories.

Most A-level textbooks are divided into chapters on each major group of hazards or hazard types. Many schools and colleges find these a useful framework for carrying out hazard research because a detailed study, for example for earthquakes, can be made of the cause, spatial distribution, responses and so on. Alternatively, centres study the unit by generalisation and then include a range of case studies appropriate to each one. However, this can be repetitive and may lead to a piecemeal study of each hazard event.

Neither way is absolutely ideal. The crucial skill, therefore, is to enmesh hazard groups with generalisations. The diagram below is a useful case-study template that you can use for each major case study that you research.

Classification	Tectonic	Climatic	Geomorphic
Type	Volcano/earthquake/tsunami	Atmosphere/hydrology	Type?
Frequency	High	Medium	Low
Magnitude	High	Medium	Low
Spatial extent	Wide	Medium	Narrow
Origin	Natural	Mixed	Anthropogenic
Generalisation 1: Causes of different types of hazard			
Physical (Regional? Local?)		Human	
• General problems caused? • How can knowledge of physical process help?			
Generalisation 2: Spatial variations in the impact of natural hazards			

Social classes/politics		Economic		Environment	
short term	long term	short term	long term	short term	long term

How do the impacts vary spatially (micro/meso/macro)?
Generalisation 3: The human response to hazards
• Type of strategy (do nothing/manage/forecast/engineering — technological fix?) • Decision makers involved? • Stakeholders and gatekeepers? • Schemes used at what scale — local? regional? • Protection/alleviation/prevention? • Hard (engineering)/soft (planning, education)?

Generalisation 4: Global trends in hazards and hazard management

- Trend?
- Becoming more frequent?
- Overall futures?
- Methods used?
- Quality of communications?
- Lessons learned?

- Link with economic development?
- Role of prediction?
- Management/protection/prevention?
- Length of warning/precursive signs?
- Orderliness of response by public?

Case-study summary — key facts

The assessment

Make yourself familiar with what is required. A good way to study is to mark, with your teacher and classmates, some examples of essays using the generic (DRUCQ) mark scheme. See what is required to achieve top-band marks in each section.

The DRUCQ mark scheme

You are allowed 1 hour and 20 minutes to write this *single* essay (you choose one title from two). You should spend about 10 minutes planning what you are going to write. Most students manage around five sides of writing, i.e. between 1500 and 2000 words. However, so much depends on your style of writing — and the breadth and depth of your research.

The essay is marked out of 60, according to the DRUCQ criteria.

'D' stands for define

Introducing, defining and describing the question, problem or issue, and identifying the data/information required to answer it (10 marks)

Marks	Criteria
8–10	A clear statement of the question, problem or issue, which shows understanding of the nature of the data needed and why. Refers to a potential range of scales and/or locations. Accurate definition of terms.
6–7	The candidate refers to the question, problem or issue raised and is able to identify an appropriate framework to approach it. Key terms are generally defined, but with a varying degree of completeness. A range of scales and/or locations suggested for evidence.

Marks	Criteria
3–5	Some reference to the question, problem or issue. Key words may be vaguely defined. Evidence of an attempt at a framework to introduce the question.
1–2	Makes a limited effort to define the question, problem or issue. Key words may be vaguely defined. Evidence of an attempt at a framework to introduce the question.
0	No attempt to define the question, problem, issue. Key words or terms are not defined. No evidence of a framework is given for the answer to the question.

'R' stands for research

Researching relevant sources, selecting appropriate case-study material and using this knowledge in detail (15 marks)

Marks	Criteria
13–15	A wide range of reading and research is evident. A wide range of excellent case studies is used (from a variety of locations and scales), with plenty of evidence of careful, relevant selection of material. Appropriate use of annotated maps, figures and diagrams accompany the written research and are incorporated into the answer. Research sources are evidenced.
9–12	Some useful research of resource material is evident from and cited by the candidate. A range of relevant case studies is used at a number of scales and/or locations, and there is evidence of careful selection. Annotated maps, figures and diagrams feature in the evidence base. Research sources are evidenced and include some personally researched exemplars. Diagrammatic and cartographic evidence may be used.
5–8	A range of relevant case studies is used from a variety of scales and/or locations. The selection data and evidence from these tend to be used descriptively rather than analytically. Research sources that are mainly standard sources of information are evident.
1–4	Case-study material is provided, but it may be used in a limited way or have limited appropriateness in location or range. Basic evidence of research is demonstrated, but the evidence may lack clarity or justification.
0	Case-study material is absent — most statements are generalised, with little precise identification of scale, location or significance of examples. No evidence of research sources.

'U' stands for understanding

Understanding of general concepts, case studies, attitudes and values, and the application of data and information to the question, problem or issue (15 marks)

Marks	Criteria
13–15	Candidate shows ability to organise data logically throughout and can apply research material fully to the question, problem or issue. Appreciation is shown of a range of values or perspectives about the issue or problem. Most or all data lead directly to an answer to the question and are well sequenced and explained. The result is a highly cogent, soundly conceived answer.

Marks	Criteria
9–12	Organises data to suit the question, problem or issue and is aware that the data are being used to support an answer. The command words become evident in the way data are utilised to provide a well-ordered and explained answer that shows conceptual understanding. Values and attitudes are appreciated and are embedded in the answer.
5–8	Some evidence of ability to organise material to suit the question, problem or issue. The level of description shows an appreciation that the material is relevant, although it may be generalised and partially ordered and explained simply. Single-value systems are appreciated rather than a variety of values.
1–4	Material is presented in a form that is brief and has some relevance to the question, problem or issue. There is limited detail and focus on concepts. Descriptive work, which the reader is expected to make relevant to the question. Limited appreciation of values and attitudes.
0	No evidence of organising material to suit the question, problem or issue. Material is descriptive, lacks any detail, is not used in answering the question or appears completely out of context.

'C' stands for conclusion

Drawing appropriate conclusions on the basis of evidence, and on-going evaluation (10 marks)

Marks	Criteria
8–10	A conclusion is clearly stated and is directly related to the rest of the essay. Conclusions or ideas drawn will generally draw upon evidence given in the essay and will have grasped the complexity of the question, problem or issue. The conclusion is likely to have been built up through the essay progressively. Ongoing evaluation a strong feature.
6–7	A meaningful conclusion based on the material in the essay, which is recalled selectively. Does appreciate the question and links back to the introductory intent. Shows some evaluation.
3–5	A vague conclusion stated in general terms that is related tenuously to the question. Very limited evaluation evident.
1–2	An attempt at a conclusion, which has the barest linkage to the question and the preceding evidence.
0	No conclusion or attempt to evaluate an answer to the question, or one which is thin and has no link to any evidence or research material included.

'Q' stands for quality

Quality of written communication, including the communication of knowledge, ideas and conclusions in a clear and logical order, and the use of appropriate geographical vocabulary (10 marks)

Marks	Criteria
8–10	The essay is well written with clear sense, coherence and style and with clear syntax. Organisation of material is into sequenced paragraphs that lead from one to the other; clear and logical arguments are developed as the essay progresses. Knowledge and ideas are integrated, and lead to conclusions. Terminology includes appropriate use of specialist vocabulary.

Marks	Criteria
6–7	The essay is generally written clearly. Syntax is usually clear; organisation of material is reflected in paragraphing and there is some attempt to sequence parts of the essay so that they link together. Some good use of appropriate geographical terminology evident.
3–5	The essay is written in a disjointed style, either using single-sentence paragraphs or with few paragraphs. Ideas tend to be 'whatever comes next'. Some geographical terminology used, although the language of the discipline is intermittent. Syntax is basic, with some errors of punctuation.
1–2	Very basic quality of language with frequent spelling and punctuation errors, low-level syntax and minimal ordering. Very occasional use of basic geographical terminology.
0	Basic standards of quality of written communication not met.

The generalisation

Stage 1

When the generalisation arrives about 6 weeks before the exam, revisit it in the specification, to see the angles on which questions could be set. For example, **Generalisation 4: Global trends in hazards and hazard management** includes at least three sub-generalisations. With your teacher you could, using past papers, develop a bank of around 10–12 essay titles, so that you can really explore the range of ideas across the generalisation.

> **Tip**
> Remember that if you do practise writing essays, it is unlikely that the title in the exam will be exactly the same as one you have practised, so you have to be flexible.

Stage 2

Each generalisation has a particular theme.
- **Generalisation 1** is concerned with causes of hazards. There is no substitute for drawing diagrams of the various types of plate boundary and their impact on earthquakes and volcanic eruptions or of how hurricanes form.
- **Generalisation 2** assesses the spatial variations of the impacts of hazards. You should prepare maps, for example of key multiple hazard zones.
- **Generalisation 3** relates to human responses to hazards. Conceptual frameworks, such as the Park model, are essential to avoid excessive description of death and damage. The 'modify the event', 'modify the vulnerability' and 'modify the loss' framework is a useful organising tool for tables.
- **Generalisation 4** is concerned with the trends in hazards and hazard management. You should use precise statistics of these trends to enhance your explanation.

You can, therefore, prepare appropriate case-study support material that is packaged in a way to target the chosen generalisation.

Stage 3

Look through all your case studies and sort out how you can summarise them to 'tell a story' related to the generalisation. Case-study summary cards can be useful.

Stage 4

Have a quick look again at the *Foundation for Study*. Learn key definitions and questions and start thinking about possible introductions. Always try to be well prepared, but not over-prepared. Above all, don't be blinkered — the generalisations are interlinked.

As well as summaries of case studies, you need to learn framework summaries of key points, as illustrated in the partially completed chart below.

Prevent or modify the event	Modify vulnerability	Modify the loss (accept or share losses)
Technological fix of physical processes by:	Change attitude and behaviour before and during event by:	Accepting losses, while the most passive response, is rarely acceptable. More common to share the loss by:
Prevention and environmental control: landslide management (Hong Kong); volcanic lava flow management (Heimaey); catchment management flood control (upper Severn)	Prediction and warning: tsunami-warning system	Aid:
Protection (hazard-resistant design):	Community preparedness:	Insurance:
	Land-use planning:	

Content Guidance

This section is designed to help you develop and assemble the knowledge needed to successfully answer an essay question based on the pre-released generalisation.

First, there is guidance on **How to research hazards effectively**, in terms of finding information, selecting and sorting, synthesis and analysis, recording and rewriting. This is followed by the specification content:

- **Foundation for study**
- **Generalisation 1: Causes of different types of hazard**
- **Generalisation 2: Spatial variations in the impact of natural hazards**
- **Generalisation 3: The human response to hazards**
- **Generalisation 4: Global trends in hazards and hazard management**

Throughout this Content Guidance, key terms are in **bold**. This should help you build up your own dictionary of essential terms.

How to research hazards effectively

Figure 1 outlines the process of researching hazards.

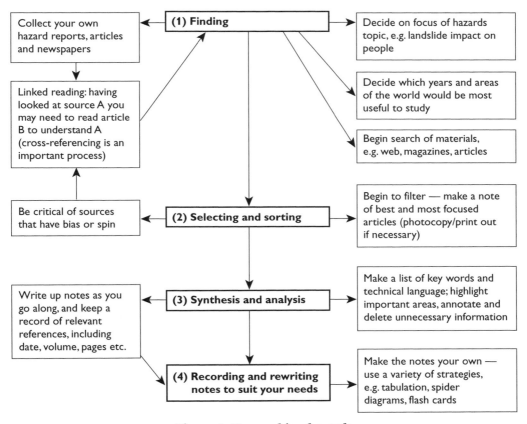

Figure 1 Researching hazards

Finding

The variety of sources

There is a mass of available information about hazards. In newspapers, the information is often text-rich, interspersed with occasional photographs. In more specialist publications, such as *New Scientist, Scientific American* and the *Economist*, information is presented in more technical language, interrupted with diagrams. Valuable information can come from listening and watching. Radio programmes (particularly those on BBC Radio 4) can be a rich source of documentary evidence. Digital satellite channels such as Discovery and BBC4 can also provide valuable, up-to-date information on hazards.

The internet connects you to a seemingly infinite source of geographical information. It is quite simple to use once a few ground rules have been established. These are discussed on p. 15.

Getting into some reading

Reading is partly about studying words and taking meaning from them. However, there are basic challenges to overcome:

- coping with *large amounts*
- trying to *understand* the difficult parts
- finding ways to *digest* and *synthesise* information
- *remembering* what you have read

It is good idea to photocopy the article/piece of writing so that you can scribble notes on it.

Coping with words

It is easy to find yourself struggling with some of the more technical or difficult words. One solution is to use a specialist geography dictionary. As you begin to synthesise information, it might be a good idea to keep a good-sized general dictionary (e.g. the *Concise Oxford Dictionary*) close to hand. It will be a useful resource when you are lost. However, often you can get a better insight into the meaning of key terms from your study texts.

You might want to develop a system for writing down words that seem important as you come across them. Consider using a card system, starting a new 'concept card' for each word/theme/idea that you think is worth the trouble (Figure 2). Keep the cards filed alphabetically so that you can find them quickly. Each time you come across the word in a different context, add the new information to the card.

Gee-gees – Global Geophysical Events (GGEs)
- What are they? Extreme natural events with very low levels of probability of occurring in any one year, but 100% probability in the much longer term.
- Examples?
  – Super-volcanoes, comets and asteroids, giant tsunamis and earthquakes.
  – Underpinned by threat of climate change.
- What's all the fuss? They are disasters that have the potential to affect the whole planet. Could also hit major metropolitan areas, leading to vast loss of life.
- The problems
  – Difficult to predict where and when.
  – Require a range of government agencies to coordinate action.
  – May not happen in the short term so why bother – cost.

Figure 2 Concept card on global geophysical events

Using the internet effectively

There is a mass of information on the internet, such as photos, blogs, text documents, video, audio (including RSS news feeds), articles, diaries and maps:

- **print** — PDF files, Microsoft Word documents and HTML web pages, online newspapers and magazines; also 'blogs'
- **audio** — RSS news feeds, specialised podcasts, BBC's 'Listen Again', Blinx website (to search for audio feeds)
- **video** — BBC's 'Creative Archive', CNN, ITN News, British Pathé (archive)
- **images** — Use a search engine to find image files and specialist hazard images on USGS and university research sites
- **maps** — Multimap, Google Local, Google Earth, 'Topozone' for a range of US maps, Geography Network

Anyone with the right hardware and know-how can post information on the internet. This means that there are no guarantees as to the quality or provenance of the information. There are also a number of information services, which may charge or be accessible by subscription only. Researching on the internet gives you the feeling of being busy but it may not be that time-productive. When using internet sources, there are some points to consider:

- Who published the information? A site maintained by a university or government organisation is probably more reliable than one maintained by a private individual.
- Who wrote the information? Material provided by a known expert in the field is likely to be reliable.
- How old is the material? If you need current statistics, check the age of the material. A site dealing with historical hazards information may not need updating as frequently as one related to news and current events.
- Why does the material exist? Many special interest groups have web pages. This does not necessarily mean that the material is biased, but it is something you should consider. There might be some reason, other than pure helpfulness, for posting information.

Selecting and sorting

The best hazards websites

There are a number of good websites on hazards. The listing below shows some of the best ones:

- **www.gesource.ac.uk/hazards/** is a high-quality, general natural hazards site. Check out the excellent maps, e.g. 'billion dollar' disaster events and world climate maps.
- The Benfield Hazard Research Centre at **www.benfieldhrc.org/index.htm** has some useful downloadable publications (e.g. *Hazard Risk* and *Science Review*) and a good newsletter.

- **http://quake.wr.usgs.gov/** is the USGS website on all matters related to earthquakes. It is a good starting point for research. Look out for general earthquake information to download some good material.
- The Edinburgh Earth Observatory at **www.geo.ed.ac.uk/quakeexe/quakes** is another fine earthquake website, with good maps of current earthquake activity.
- The European Strong Motion Database is a rich source of information. Find it at **www.isesd.cv.ic.ac.uk**. (You have to register, but the data are good.)
- **http://volcano.und.edu/** takes you to the Volcano World homepage. There is a feast of information for you to check out, including up-to-date eruptions.
- **www.ess.washington.edu/tsunami/index.html** is a general tsunami site hosted by the University of Washington. Find the link to the animations.
- **www.nhc.noaa.gov** is the website of the US Hurricane Centre. It is probably the best general climate site for hurricanes and tornadoes. It has current watches and warnings.
- The Dartmouth Flood Observatory at **www.dartmouth.edu/~floods/** has maps and tables — both historic and up to date — for all major global flood events.
- **www.nwl.ac.uk/ih/nrfa/** is the website of the National Rivers Flow Archive. You can download detailed daily river data for UK rivers.
- The USGS website also has a selection of hazards fact sheets. See the link at **http://water.usgs.gov/wid/index-hazards.html**
- The International Red Cross homepage is at **www.ifrc.org/**. World disaster reports can be obtained here.

Developing a hazards diary

You could develop a hazards diary throughout the A2 year. Writing a few sentences each week about particular events will certainly help when it comes to revision. It may be an idea to include a world map in your diary and to indicate on the map when events occur. Alternatively, you could develop your own spreadsheet to log and track events (Table 1).

Table 1 Logging hazard events

Date	Location	Type	Cause	Duration and magnitude	Estimated damage	Deaths
16 Aug 2004	Boscastle, North Cornwall	Flood	Intense rainfall plus prolonged storm (Bodmin Moor) plus rising tide	Short ($< \frac{1}{2}$ day); 180 cumecs (estimated)	Localised flooding of homes (60) and village; damage to infrastructure; cars in harbour	None

The diary approach is really important for the 'big ones' (e.g. Katrina, Asian tsunami), for which information is gradually released over the months following the event (Figure 3).

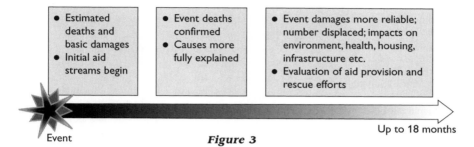

Figure 3

Synthesis and analysis

Making sense of text

To get to grips with ideas in texts and 'make them your own', there is really no alternative to unpacking, distilling and synthesising information from the documents you have collected. This can be a real hatchet job that includes crossing out, cutting off and throwing out parts of the prose. You will be forced to grapple with the text as you read, deciding what is particularly important, what is moderately important and what you can do without.

Synthesising text

Distilling and refining longer articles and prose are important skills. It could be that you are intending to use the material as a brief case study. The original writing may be in the form of a newspaper or magazine article; it could be a series of linked web pages or a PDF document you have downloaded. The first stage is to work out which parts are important and delete or ignore the remainder. Then start to make sense of the remaining portion.

Example

Establish the bare facts
The date and location of the incident are important (i.e. North Yorkshire, village of Lower Hawnby, River Rye; 19 June 2005; flooding).

North Yorkshire villagers are determined to have a better New Year after their lives were devastated by the most powerful flash floods in living memory.

From the top of the village of Hawnby you can see for miles across the North York Moors. It nestles in a picturesque valley through which the River Rye gently flows.

But it was that setting which proved disastrous for the village, along with parts of Helmsley, the nearest market town,

Context of incident

Three inches (approximately 75 mm) of rain fell in 1 hour. The impact of this high-intensity storm was worsened by:

- the preceding conditions — high river levels and hard-baked ground created a semi-impermeable surface
- the hilly catchment — probably increasing the steepness of the storm hydrograph

Impacts of the flood incident

The impact of the flood was worsened by Church Bridge acting as a dam, resulting in a catastrophic release of water.

Short-term impacts include:

- village cut off
- loss of livestock
- three dogs lost from kennels
- shocking and harrowing experience for local people
- 40 footbridges swept away

Longer-term impacts include:

- loss of fields
- access only by temporary bridges — inconvenience
- loss of passing trade, particularly from walkers
- council costs of £5 million; National Park costs of £600 000 for footpath repairs

and several other villages on the fringes of the moors when on the night of 19 June last year, 3 inches of rain, the equivalent of last month's normal rainfall, fell in just 1 hour with catastrophic consequences.

It had been a hot day — too hot to sit outdoors according to Sonia Leeming, who with her husband, Darren, runs the village store, post office and tearooms in Lower Hawnby. The thunderstorm which followed the heat wave was not unexpected, nor the power-cut. 'But by the time I'd fired up the generator, a neighbour had rung to say water was running through their house,' recalls Darren.

Much worse was to follow — millions of gallons of surface water swept down from the hard-baked hillside into the already raging river. The normally placid waterway roared down the valley, sweeping away all in its path, from bridges to giant oak trees. While some stock was rescued, other sheep and cattle drowned and the sight of their bodies floating in the torrent was especially upsetting to the shocked villagers.

'It had more or less stopped raining and it wasn't fazing us at all, but suddenly all this water was just there,' says Darren. 'Apparently Church Bridge had been acting as a dam; when that went, all the water was released. It sounded like a jet engine on full bore — it was so loud, we couldn't hear the thunder any more.'

Alongside the river, Ray Howard and Mary Griffiths suffered most. They run the local boarding kennels. On that night, they managed to rescue 27 dogs but lost three. Six months later the couple face having both their home and kennels rebuilt, although the kennels are reopened for business. As they face the New Year, a party for the villagers paid for by the local Coop boosted morale.

The catastrophic effects of those few hours of torrential rain are still evident in the huge trees piled-up the sides of the valley leading to Hawnby, in the church still being dried out, in fields put out of action by a layer of mud, but mainly in the miles of ruined footpaths in the area.

Hawnby still suffers inconvenience; for days afterwards the village was completely cut off and even now the only access is by temporary bridges, and suppliers of goods have to come the long way round, while what were formerly short journeys to playgroup or school now become major operations.

Local businesses that rely heavily on walkers who explore the network of footpaths have suffered a loss of trade in the months since the floods. 'But we are still here, and welcoming visitors.'

About 40 footbridges were washed away as well as road bridges. North Yorkshire County Council faces a £5m bill for repair and replacement. The National Park Authority is to spend £619 000 on footpath and bridleway repairs.

The future?

Given the location of the village, there is a real threat of further flooding. This may be amplified by impacts of climate change. There may be an increased incidence of flash floods, which are harder to predict (Environment Agency).

Meanwhile, hanging over the villagers is the threat of more flooding. The chairman of the Environment Agency, Sir John Harman, has warned that climate change could lead to further torrential thunderstorms and flash floods, the location of which will be hard to predict.

Adapted from the *Dalesman*, January 2006

Working with diagrams

There are various types of diagram that can be used to illustrate hazards information:

- **Pictures** or **pictorial diagrams** that attempt to represent the essential features of something, such as a diagram of a modification to a building designed to make it more earthquake-resistant. These simple diagrams may also include photographs.
- Diagrams that depict **interrelationships** between ideas, processes and concepts using words, lines, circles, arrows and/or boxes.
- **Graphical diagrams**, including maps, charts and graphs. These often display spatial information or changes over time (see Figure 4).

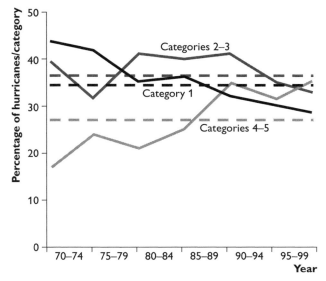

Figure 4 Development of cyclones of different intensities, 1970–99
(dotted lines indicate averages)

Tips for interpreting charts and graphs

- Take time to get a 'feel' for the purpose of the chart or graph.
- Start by looking at the title, scales and axes. Identify units.
- Study one or two points and make sure that they make sense.

- Look at the overall shape of the curve(s) — it might be familiar, for example linear, exponential, bell-shaped or skewed.
- Look for patterns, such as peaks and troughs. What might be causing such relationships? You should also look for anomalies.
- Examine the source of the data and think about how the information is presented. Think carefully about validity before drawing conclusions.

Diagrams are often used in books and articles to summarise complex situations. By making the reader think carefully about the individual components and their connections, they may also provide new insights into a situation.

Handling tabular information

Much useful information can be presented in a table. Tables vary in form and complexity, depending on sources; they may include text, data and even images. Tables are normally used to summarise a large amount of information. Interpretation of tables takes time and is an acquired skill. Start by making a basic analysis:

- What are the obvious trends and patterns? Look at any end-of-row summaries.
- Are there anomalous data/entries? Scan rows and columns.
- Begin to break down the information into simple units and use percentages.

This is illustrated in Figure 5 (p. 21).

Handling controversial issues

Hazards geography throws up controversial issues, for example:
- Natural hazards are becoming increasingly frequent.
- Climate change is making hurricanes stronger.
- Earthquakes are predictable hazards.
- Deforestation is leading to more flash flooding.

The way to handle controversial issues is to be able to recognise different sides of the story and to acknowledge different opinions, as in the following example, about deforestation:

- 'Logging does not raise flood risk' (BBC News, 2006)
- 'Flood tragedy in Thailand linked to deforestation' (WSWS.org, 2001)
- 'Environmentalists have warned that deforestation on Java is exacerbating the effects of floods and landslides' (BBC News, 2006)
- 'To generally conclude that deforestation in the Philippines is not a cause of flooding is misleading' (Centre for International Forestry Research, 2005)

There may be several explanations and reasons. It is a good idea to keep records of sources and dates so that you can quote them in an exam. Acknowledging the complexity of these contentious issues is likely to be well rewarded. Natural hazards stem from a complex mix of factors, including climate change, socioeconomic factors that cause people to live in risk areas and inadequate hazard preparedness. 'Globalisation' of reporting means that we are now much more aware of hazards.

Floods responsible for greatest number of deaths in Americas

Most deaths from droughts and earthquakes

Earthquakes and tsunamis far more hazardous in medium developed countries; also caused the most deaths for any single category

Overall, forest fires and volcanoes least hazardous in terms of number of deaths

Volcanic eruptions (globally) create fewest deaths at only 313; may be due to improved forecasting and evacuation strategies; also shows the relative rarity of large events

Windstorms create surprising number of deaths globally — over 31000

Most developed countries suffer fewest deaths (only about 6%); medium developed suffer 57% and least developed suffer about 37% of deaths

Oceania suffers fewest deaths; may be function of low exposure to hazards, but also low population densities averaging 3.7/ha

In Europe, most deaths are due to extreme temperatures (e.g. France 2002)

Total number of people reported killed in natural disasters, by type of phenomenon and by continent, 1995–2004 (extract from World Disaster report, 2005)

	Africa	Americas	Asia	Europe	Oceania	Most developed	Medium development	Least developed	Total
Avalanche/landslide	251	1742	6219	416	128	370	7576	810	8756
Drought/famine	4551	59	270923	n.a.	88	n.a.	927	274694	275621
Earthquakes/tsunami	3114	2290	292050	20727	2200	8231	304237	8761	321229
Extreme temperatures	200	2325	9817	32888	1	32986	11378	867	45231
Floods	9176	38000	44219	1414	33	3599	76093	13150	92842
Forest fires	114	83	125	107	9	146	288	4	438
Volcanic eruptions	254	52	3	n.a.	4	20	39	254	313
Windstorms	1385	25271	33958	720	250	4514	52391	4679	31584
Other natural disasters	n.a.	3	448	n.a.	n.a.	n.a.	451	n.a.	451
Hydrometeorological totals	15677	67483	365709	35545	509	41615	149104	294204	484923
Geophysical totals	3368	3042	292053	20727	2204	8251	304276	9015	321542
Totals	**19045**	**70525**	**657762**	**56272**	**2713**	**49866**	**453380**	**303219**	**806465**

Figure 5 Interpreting information given in tables

Table analysis — general points
- In 1995–2004, biggest killers were floods in Americas (46%) and earthquakes/tsunami in Asia (42%) and Oceania (67%)
- Hydrometeorological events are bigger killers than geophysical (60%:40%)
- Medium developed countries suffered most deaths overall (57% of total)
- In Europe, 51% of deaths related to extreme temperatures

Recording and rewriting notes

Having collected and synthesised information from a variety of sources, the next task is to make these your own. Think of these new notes as a *handier version* of the original material. The products of synthesis and analysis should not be a shortened rewrite of the original piece. You should attempt to present the information in a different way — for example, tabulation (tables and matrices), diagrams (spider diagrams, mind maps, simple flow diagrams) and bulleted or numbered lists. These can include a rewrite, reconstruction and reworking of your original notes and ideas from the initial synthesis and analysis.

Points to bear in mind

- Making notes and reworking is not a single skill. It should include a range of activities that keep you interested.
- You are writing for yourself, so don't worry too much about neatness or explaining points that you find obvious.
- Reworking notes is more strategy than skill. You should be constantly thinking — why is this useful and how can I rework the information more clearly?

Creating tables

The golden rule is *deciding what you want to communicate* about the information and data in a table. Always give the table a title (even in your own notes) and reference the source of the data.

Hints on constructing a table:
- Decide what the table is for. What will the rows and columns be about?
- Plan the layout on the page. How many rows and columns will be required? To make the meaning clearer, you may need to construct the page in landscape format.
- Enter the data/text into the table, adjusting the table properties if necessary. Numerical data should have the appropriate units in the column headings.
- If the table is large, with lots of figures and text, check that it all makes sense. Do this by scanning the rows and columns, looking for items that are out of place.
- Remember that tables can be kept simple — for example, matrices of advantages or disadvantages, or LEDC versus MEDC.

Spray/spider diagrams and mind maps

Making your own diagrams is a good idea. Used correctly in an exam, maps and diagrams will help you to express yourself concisely and can save writing time. They offer an opportunity to increase your depth of learning about a subject and to analyse and express that new understanding to the examiner. Diagrams are also useful as revision tools and can be used to summarise large amounts of text.

- In the centre of a piece of A4 paper, identify the mind-map topic.

- Draw the main topic branches. These need to contain the main categories of information that are to be included in your spider diagram, such as answers to What? When? Why? Where? So what?
- Add any sub-categories, e.g. short- and long-term effects.

The example in Figure 6 is a case study of the eruption of Mount Pinatubo, in the Philippines:

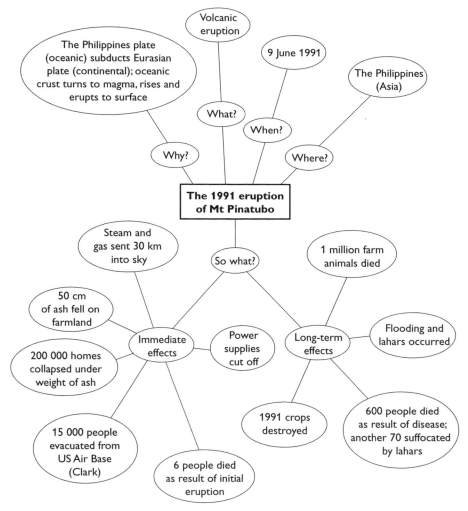

Figure 6 *Mind-mapping a case study*

Auditing your research information

Once you have collected and reworked some articles, you should keep track of them. One way is to create a table to summarise them (see Figure 7, p. 24).

Key:
- O Some linkage
- ● Strong linkage
- HDI Human development index

Case study number/ research paper: date and details	Causes of tectonic, meteorological, climatic and geomorphic hazards								Level of development			Human response and future trends			
	Earthquake	Volcano	Tsunami	Hurricane	Drought/flood	Rockfall/landslide	Avalanche	Multiple hazards	Most developed/high HDI	Medium development/HDI	Lowest development/low HDI	'Do nothing' strategy	Avoidance/education	Hazard prediction	Becoming more frequent?
1 Drought in France, 2002					●				●			O	O	O	O
2 Earthquake in Gujarat, India, 2001	●									●					
3 Landslide in Po Shan Hong Kong, 1972						●		●	●				●	●	O
4 Floods in Carlisle, UK, 2005					●				●				●	●	O
5 Drought/flood in Ethiopia, 2006					●			●			●	O	O		O
6 Earthquake/tsunami in Java, Indonesia, 2006	●	●								●			O	●	

Figure 7 Summarising your research

The table approach is particularly useful because it gives an idea of spatial coverage. In an exam, it is always best to try to use examples from contrasting parts of the world, as well as regions with different levels of development.

An alternative to the table is to use an A3 world map, marking on it the locations of case studies you have researched. Use an appropriate key for hazards such as earthquakes, volcanoes, floods, hurricanes/cyclones, and use a subscripted number adjacent to each hazard as a reference to a particular research document.

Foundation for study

It is essential to learn the key concepts. The **Foundation for study** provides knowledge and understanding of:
- **key terminology**, such as the difference between natural and environmental hazards or a hazard and a disaster.

- ways of **classifying natural hazards**. Questions often require you to select your case studies from a particular hazard group (e.g. tectonic hazards), so you must be clear about the groupings.
- how to profile hazard **types** and hazard **events**. This can be combined with a spreadsheet analysis, so case studies can be compared and put in a wider context.
- the importance of **risk assessment** and vulnerability as a means of studying how hazards can be prevented from becoming disasters.

Key definitions

A **natural hazard** can be defined as 'a perceived natural/geographical event which has the potential to threaten both life and property' (Whittow). The geophysical event would not be hazardous without, for example, human occupancy of its location, i.e. river or coastal floods would not be hazards if people did not live in river floodplains and coastal plains. Since hazards occur at the interface between natural and human systems, it is unlikely that any hazard is truly natural. There is a continuum from natural to quasi-natural to man-made (see Figure 8). There is mounting evidence that worldwide environmental changes, particularly those associated with the enhanced greenhouse effect, will produce large-area hazards (context hazards) as opposed to site-specific hazards, because they will exacerbate atmospheric hazards (e.g. storms/floods) and facilitate the spread of diseases (e.g. malaria).

An **environmental hazard** is defined as 'the threat potential posed to humans or nature by events originating in, or transmitted by, the natural or built environment' (Kates). This encompasses a wide spectrum (see Figure 8).

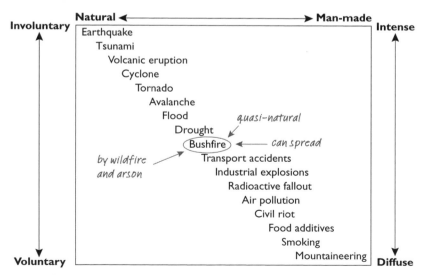

Figure 8 A spectrum of environmental hazards

The terms 'hazard' and 'disaster' are often used casually or synonymously. Figure 9 summarises how a disaster is the realisation of a hazard, when it 'causes a significant impact on a vulnerable population'. The main issue is to establish a threshold for significant impact in terms of numbers killed or affected.

(a) No hazard or disaster

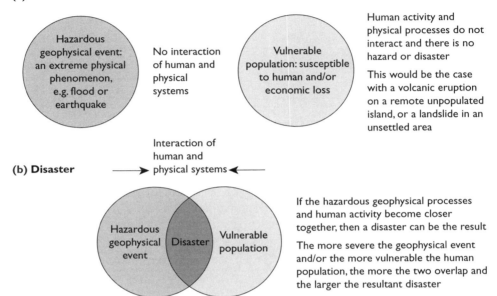

Figure 9   The relationship between hazard, disaster and human vulnerability

For inclusion in the Centre for the Reporting of Epidemiology of Disasters (CRED) Emergency Events Database, a disaster must have:

- killed ten or more people
- affected at least 100 people (for drought or famine, 2000 people have to be affected)

An appeal for international assistance or a national government disaster declaration takes precedence over the above two criteria.

There have also been attempts to include a damage threshold based on 1% of GDP lost. However, fluctuating currency values have made this difficult to manage.

Classifying hazards

Hazards have been classified using a wide range of criteria and approaches, including the nature of the geophysical process/cause, magnitude and frequency, duration of impact, warning time and spatial distribution.

The purposes of classification include:
- risk assessment
- understanding spatial patterns
- understanding how hazards might impact on people
- helping to manage responses to hazards
- the understanding of processes and their interrelationships

As classification puts hazard events and types into a wider framework, the forces of simplicity (ease of use) and accuracy (scientific precision) are often diametrically opposed. Any classification needs to be fit-for-purpose for the user group, be they researchers, planners, insurance companies or hazard managers.

Methods of classification: by origin

The most widespread classification of hazards is by cause or origin. Figure 10 shows a standard tripartite division.
- **Geological hazards** originate within the Earth and are related to internal constant processes, such as those caused by plate tectonics.
- **Climatic hazards** are caused by extreme fluctuations in atmospheric processes.
- **Geomorphological hazards** are those relating to surface processes, particularly land instability.

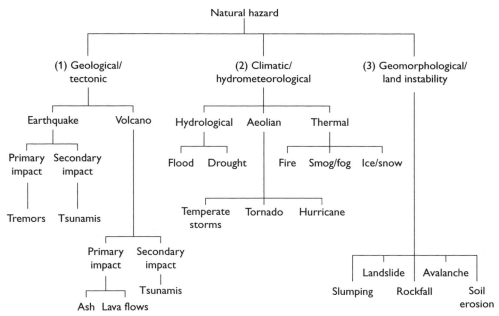

Figure 10 Natural hazards classified by origin

An **avalanche** could be classified as either an instability hazard *or* a snow–ice hazard.

In Figure 11, hazards are classified according to the degree of human impact on the cause.

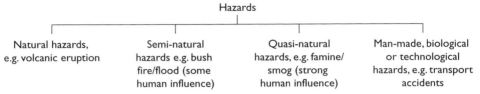

Figure 11 Human impact on causes of natural hazards

Figure 11 includes:
- quasi-natural hazards (e.g. smog and desertification), which often result from inter-action between natural and human processes
- biological hazards (e.g. epidemics), which affect living organisms and result from biological processes

Methods of classification: by impact

There is a range of classifications based on impacts — for example, spatial occurrence, scale of impact, duration of impact and speed of onset. These are all features of a hazard profile (p. 29).

One classification of impacts uses a 'severity of impact score' based on the loss of life and number of people affected. A located choropleth shading system (Figure 12) could be used to classify the impacts of hazards to create a **scale of severity** (up to 8 by adding deaths and numbers affected).

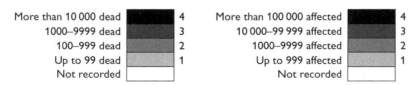

Figure 12 Scales of severity

Alternatively, established scales can be used to classify the severity (**magnitude**) of the physical processes:
- **Earthquake magnitude** is measured using the **Richter** scale, which is logarithmic. This means that magnitude 9 is ten times greater than magnitude 8. This would have a major impact on the areal extent of damage and the length of time the shaking took place. The **Mercalli** scale, which uses a series of damage descriptions, is also used to work out the intensity of shaking.
- **Volcanic eruptions** are classified using a range of criteria. The volcanic explosivity index (scale 0–8) includes data on eruption rate, volume of ejected material, height of eruption column and duration of continuous blasts in hours.
- **Hurricane** disaster potential is measured by the **Saffir–Simpson** scale, which ranges from 1 to 5. A category 5 catastrophic hurricane is likely to cause 250 times

more damage than a category 1. Emergency services and governments use the scale to predict the disaster potential in areas of risk.

- **Tornadoes** are classified by their path width and length, and wind speed. The **Fujita intensity** scale used in the USA is based on the damage caused; the **TORRO intensity** scale (T0–T10 super tornado) is based on the Beaufort wind-speed scale and observed damage.

There is an **inverse relationship** between frequency and magnitude, i.e. 'mega' events of high magnitude are rare.

Other hazard events are classified by calculating the **recurrence interval**. This is the calculated **frequency** at which the event (e.g. flood or storm surge) is likely to recur. For example, the Carlisle floods were calculated to have a recurrence level of 170 years, i.e. they were a 1-in-170-year event.

Hazard profiling

An overall picture of a hazardous event can be obtained by drawing a **hazard profile**. This shows the relative dimensions of the hazard type by a plot of its characteristics, according to their positions on six scales representing six indicators (Figure 13).

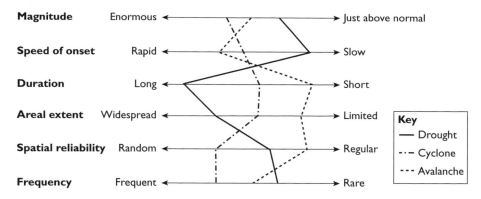

Figure 13 Hazard profiles for a drought, a tropical cyclone and an avalanche

The indicators are:
- **magnitude** — the size of the event; for example, force 10 on the Beaufort scale for wind speed, the maximum height or discharge of a flood or the size of an earthquake on the Richter scale.
- **speed of onset** — this is rather like the 'time lag' in a flood hydrograph. It is the time difference between the start of the event and the peak and ranges from rapid events (e.g. the Kobe earthquake) to slow events (e.g. drought in the Sahel of Africa).
- **duration** — the length of time that the environmental hazard exists. This varies from a matter of hours (urban smog) to decades (drought).

- **areal extent** — the size of the area covered by the hazard. It can range from small scale (an avalanche chute) to continental (drought).
- **spatial reliability** — the distribution of the hazard; whether it occurs in particular locations (e.g. plate boundaries) or is widely dispersed across the world.
- **frequency** — how often an event of a certain size occurs. For example, a flood of 1 m in height may occur, on average, every year on a particular river; a flood of 2 m in height might occur only every 10 years. The frequency is sometimes called the recurrence interval.

Hazard profiles are useful when using a spreadsheet to compare hazards and the occurrence of events. It may be possible to use a definite scale for plotting particular features to enable effective comparison (Figure 14). Extra criteria, such as death toll and damage in US dollars, could be added.

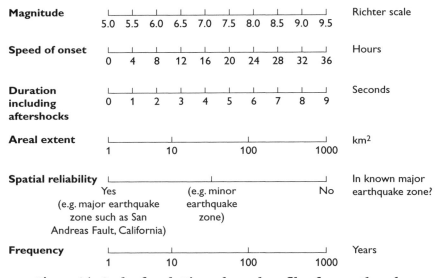

Figure 14 Scales for plotting a hazard profile of an earthquake

The importance of risk assessment

Risk can be defined as the exposure of people to a hazardous event that may have a short warning time. It might be assumed that risk to hazard exposure is involuntary — in reality, people may consciously place themselves at risk. Reasons for this include the following:

- Hazards are unpredictable. It is difficult to know when or where an event will occur and what the magnitude of the event will be.
- Natural hazards vary in space as well as through time because of changing physical factors and human activities. For example, a semi-extinct volcano such as Mount St Helens was not expected to cause a major eruption. With the rise of sea level,

places that were once safe are now at risk: low-lying coastal plains are more prone to storm surges and floods. Deforestation of watersheds could lead to less interception and more flashy hydrographs, thus increasing the frequency and magnitude of flood events.

- People stay in hazardous locations because of a lack of alternatives. The most vulnerable and poverty-stricken people are often forced to live in unsafe locations (e.g. hillsides, floodplains and regions subject to drought). There may be economic reasons linked to their livelihoods, such as subsistence farming. A shortage of land, or a lack of knowledge of alternatives, may promote stability as it is never easy to uproot and 'risk' a move to another location.
- Many people subconsciously weigh up the benefits versus the costs. Fertile farming land on the flanks of a basaltic volcano or on alluvium-covered floodplains or the attractive Californian climate outweigh the risks from floods, eruptions or earthquakes.
- Perceptions of hazard risks tend to be optimistic. Individuals accept that hazards are part of everyday life or result from 'acts of God'. They also have confidence in the technological fix. They seek solace in statistics, such as those published in the USA that show that each year only 1.8% of US households are affected by floods, hurricanes, earthquakes or tornadoes/windstorms and that the risk of death is far lower than that from influenza or car accidents. People also perceive regularity in irregular events. They may speculate that once a hazard event has occurred, they will be safe for the next few years.

Figure 15 outlines the risk perception process.

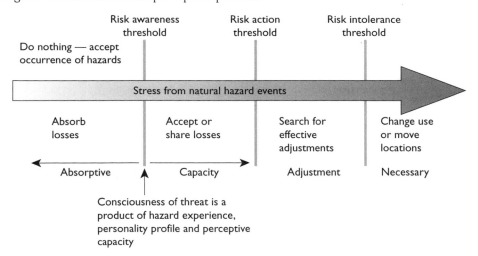

Figure 15 The risk perception process

Risk assessment and management

Risk assessment defines the likelihood of harm and damage. Figure 16 shows that **risk communication** takes place between the population at risk, risk assessors and policy makers to reach an acceptable decision. This could be either to live with the risk and educate and train people to cope, or to treat the risk by mitigation or prevention. Effective integrated risk management strategies are developed that involve the consideration of social, economic and political factors.

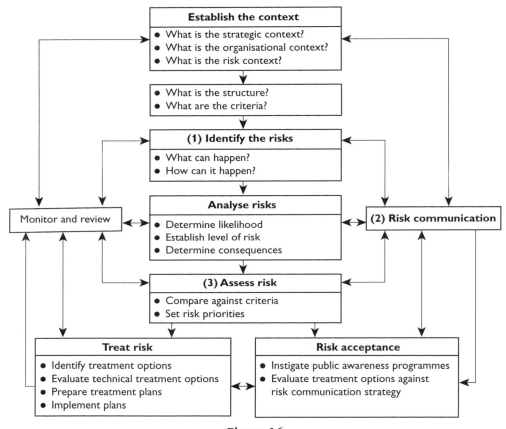

Figure 16

People's wealth (capital) and level of technical expertise affect the degree of resilience of vulnerability of the population (p. 76). Managing vulnerability and modifying the loss prevent hazard events from becoming disasters.

Review

- The *Foundation for Study* unit enables you to distinguish between hazard and disaster.
- It shows you how to classify and profile hazards.
- It emphasises the importance of risk analysis, assessment and management in order to make individuals, groups and societies less vulnerable to hazards.

Generalisation 1: Causes of different types of hazard

Earthquakes and volcanic activity are brought about by large-scale tectonic processes and are, therefore, entirely natural in origin. Other hazards in the physical world may not be entirely natural. These are sometimes called **modified physical hazards**. The role of humans as the catalyst for the increasing incidence and severity of some natural hazards is beginning to be appreciated (Figure 17). This is particularly true in the case of river flooding. Intense precipitation is the precursor to a flood event, which can be made worse by building on the floodplain and deforestation of the upper slopes, for example.

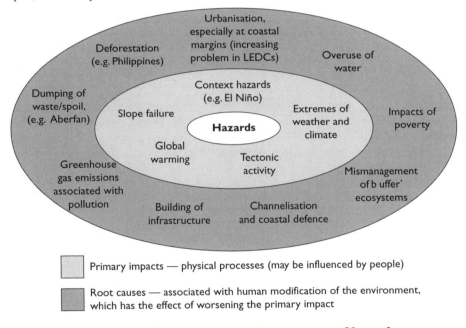

Primary impacts — physical processes (may be influenced by people)

Root causes — associated with human modification of the environment, which has the effect of worsening the primary impact

Figure 17 Primary impacts and root causes of hazards

Geological hazards

The three main geological hazards are earthquakes, volcanoes and tsunamis. Earthquakes and volcanoes release vast quantities of energy in a short time, often quite suddenly. Tsunamis are secondary hazards that result from earthquakes near or under the oceans. It is the surface manifestations of geological hazards that have the greatest impact.

The role of plate tectonics and the geography of tectonic activity

The distribution of earthquakes and volcanoes (Figure 18) reveals the following patterns of tectonic activity:

- the **oceanic fracture zone (OFZ)** — a belt of activity through the oceans along the mid-ocean ridges, coming ashore in Africa, the Red Sea, the Dead Sea rift and California
- the **continental fracture zone (CFZ)** — a belt of activity following the mountain ranges from Spain, via the Alps to the Middle East, the Himalayas to the East Indies and then circumscribing the Pacific
- isolated pockets of volcanic activity (e.g. Hawaii) over localised **hot spots**
- scattered earthquakes in continental interiors associated with volcanic islands

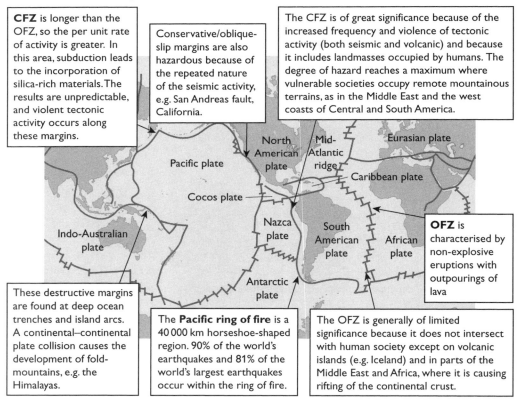

CFZ is longer than the OFZ, so the per unit rate of activity is greater. In this area, subduction leads to the incorporation of silica-rich materials. The results are unpredictable, and violent tectonic activity occurs along these margins.

Conservative/oblique-slip margins are also hazardous because of the repeated nature of the seismic activity, e.g. San Andreas fault, California.

The CFZ is of great significance because of the increased frequency and violence of tectonic activity (both seismic and volcanic) and because it includes landmasses occupied by humans. The degree of hazard reaches a maximum where vulnerable societies occupy remote mountainous terrains, as in the Middle East and the west coasts of Central and South America.

North American plate / Mid-Atlantic ridge / Eurasian plate / Pacific plate / Caribbean plate / Cocos plate / Nazca plate / South American plate / African plate / Indo-Australian plate / Antarctic plate

OFZ is characterised by non-explosive eruptions with outpourings of lava

These destructive margins are found at deep ocean trenches and island arcs. A continental–continental plate collision causes the development of fold-mountains, e.g. the Himalayas.

The **Pacific ring of fire** is a 40 000 km horseshoe-shaped region. 90% of the world's earthquakes and 81% of the world's largest earthquakes occur within the ring of fire.

The OFZ is generally of limited significance because it does not intersect with human society except on volcanic islands (e.g. Iceland) and in parts of the Middle East and Africa, where it is causing rifting of the continental crust.

Figure 18 The geography of tectonic activity

The two belts demarcate the boundaries of the tectonics plates. Motion of these plates results in significant tectonic activity along their margins, with much lower levels of activity within their interiors. There are three types of plate boundary:

- **conservative** margins (oblique slip or sliding) — one plate slides against another (rare, but well developed in the San Andreas fault in California)
- **divergent/constructive** margins — mostly displayed in the mid-ocean ridge systems of the OFZ
- **destructive** (convergent or subductive) margins — well developed along the CFZ

Variations in scale and intensity at the plate margins

Variations in tectonic activity at the plate margins result from the following:

- The OFZ is longer than the CFZ. For all the new crust created along the OFZ in a year, an equivalent amount has to be consumed along the CFZ by subduction. Therefore, the per unit rate of activity is greater along the CFZ.
- The OFZ is characterised by the upwelling of magma. The rock is hot, fluid and plastic, with low rigidity. Seismic shocks tend to be smaller.
- Along the CFZ, the rocks of the lithosphere are cold and rigid. Subduction causes much shearing, so earthquakes are frequent and generally of large magnitude. All the 'biggest' earthquakes are distributed along the CFZ.
- Volcanic activity is associated with both constructive and destructive margins, but differs in character:
  - Along the OFZ, it is basaltic and non-explosive. Eruptions are 'quiet' and characterised by outpourings of lava, as in Iceland. The same is true of oceanic hot-spot volcanoes, such as those of Hawaii.
  - Along the CFZ, subduction leads to the incorporation of silica-rich 'continental' rocks through remelting, which changes the magma composition to inter-mediate or acid. This results in much more unpredictable and violent eruptions, generating more pyroclastic material.

Distribution of earthquakes and volcanoes

Earthquake distribution

This tends to be tightly focused in linear bands (belts) but can occasionally occur elsewhere, for example the intra-plate earthquakes in New Madrid, USA (1811–12) and Dudley, 2002 (both examples of intra-plate earthquakes). The most intense and frequent earthquakes are associated with subduction/destructive zones and sometimes active conservative margins, for example San Andreas, USA. High-inten-sity shake events are *not* normally associated with non-active constructive (trans-form) boundaries.

Volcano distribution

A high proportion of volcanic activity occurs at or near plate margins (inter-plate). It can be at constructive margins (e.g. Iceland) or destructive (e.g. Andean volcanoes in Chile and Peru). Generally, distribution is both linear and clustered. Continental margins are associated with the most violent volcanic eruptions (owing to the presence of both oceanic and continental crust). Volcanoes also occur intra-plate at hot spots (e.g. Hawaii in the Pacific).

Exploring meteorological and climatic hazards

Almost 75% of disasters are climate-related and occur disproportionately in developing countries. In countries with the highest levels of development, the *absolute* economic damage is invariably highest, but the *relative* damage can be worse in the medium and lowest developed countries.

Understanding the distribution of hurricanes and tropical cyclones

Hurricanes and cyclones are predictable in terms of their spatial distribution (see Figure 19). They are concentrated in the tropics, more specifically between 5° and 20° north and south of the equator, but some hurricanes are now occurring outside these traditional spawning grounds (e.g. Catarina, southeast Brazil, 2004). Once generated, they tend to move westwards and be at their most destructive. Areas most associated with them include:

- the Caribbean/Gulf of Mexico and the western side of Central America — 28% of total ('hurricanes')
- the Arabian Sea/Bay of Bengal area — 8% of the total ('tropical cyclones')
- southeast Asia and Madagascar — 43% of the total ('typhoons')
- northern Australia — 20% of the total ('willy-willies')

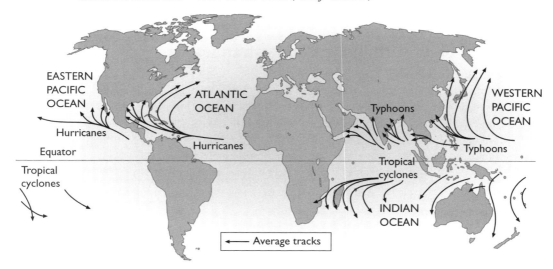

Figure 19   Distribution of hurricanes and tropical cyclones

For a hurricane to develop:
- it must be in an oceanic location with a sea temperature over 26.5°C

- the location must be at least 5° north or south of the equator, so that the Coriolis effect can bring about the maximum rotation of air. (The Coriolis 'spinning' effect is zero at the equator and increases towards higher latitudes.)
- rapidly rising moist air (from the warm sea) cools and condenses, releasing latent heat energy which fuels the storm. Hurricanes fade and die over land as this source is removed.
- low-level convergence of air occurs in the lower circulation system. This, the inter-tropical convergence zone (ITCZ), is thought to be the precursor of tropical storms.

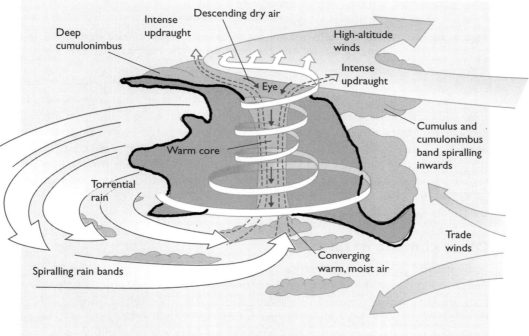

Figure 20 Evolution of a hurricane

Impact of climate change on tropical cyclones

There is a controversy over whether global warming is resulting in hurricanes being more numerous and more severe. The increase in the incidence and intensity of hurricanes has been attributed to the natural longer-term climatic cycle. Many scientists now think that a surge in sea temperatures since 1970 has already made hurricanes more severe. In particular, they have:

- been more intense
- been of longer duration
- involved more precipitation
- followed a less predictable path

However, the frequency either has been reduced or remains unchanged.

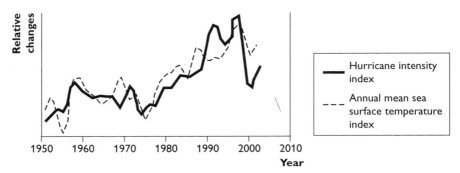

Figure 21 Hurricane intensity versus sea temperature (combined figures for the north Atlantic and the western north Pacific)

Figure 21 seems to show evidence that there is a strong correlation between higher sea-surface temperatures and more intense storms. However, other scientists argue that it is impossible to know whether recent hurricanes have been more intense because of global warming — a much longer record is required. Most researchers are agreed that there is a 'human fingerprint' in hurricane trends. This might change both the nature of hurricanes and their distribution, possibly into areas with cooler waters that were not previously considered spawning grounds. There is also a pronounced oscillation between severe hurricanes in the Atlantic (2005) and severe typhoons in the Pacific (2006).

Other climatic hazards

Mid-latitude (temperate) storms

These storms affect large areas of the world, but become hazardous only when they affect densely populated areas such as Europe. Although not normally life-threatening, they can be costly and cause widespread damage. They may also act as precursors to other hazardous events such as landslides and localised flooding.

Recent research (Brayshaw 2005) has suggested that climate change may have an impact on such storms:

- Storm tracks will shift poleward.
- The total frequency of storms will decrease.
- The frequency of intense storms will rise (the increased water vapour in the atmosphere provides an additional source of latent heat).

These impacts may also vary spatially:

- **Mid-latitude southern hemisphere**   There may be severe consequences for water supplies in southern Africa, southern Australia and New Zealand, because a large part of the precipitation here is related to storms.
- **The Mediterranean and Middle East regions**   There may be decreased precipitation, which may result in decreased glaciation in mountainous regions of central and southern Europe.

- **Northeast Atlantic and northwest Europe**   Studies predict increasing storm activity. There may also be more extreme wind events.

Tornadoes

A **tornado** is a rapidly rotating column of air extending downwards from a cumulonimbus cloud. Tornadoes are small and short-lived. However, they concentrate energy and are capable of doing enormous damage

Tornadoes occur on all continents of the world except Antarctica. The Great Plains of the USA account for 75%. 'Tornado alley' stretches through Texas, Oklahoma, Kansas, Arkansas and other central states (Figure 22). The concentration in the USA is due to the meeting of warm, moist air from the Gulf of Mexico with colder, drier air from the Arctic north.

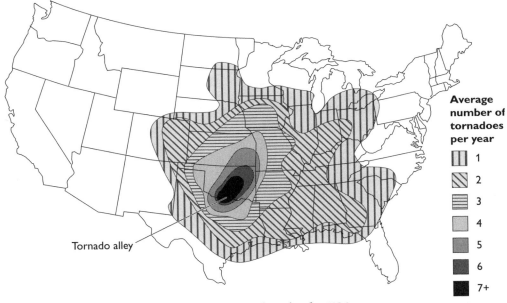

Figure 22 Tornadoes in the USA

On average, over a 30-year period, 33 tornadoes are reported each year in the UK. In reality, the actual annual figures can vary dramatically. These tornadoes tend to be localised, with limited damage. Recent examples include the one that occurred in Birmingham (July 2005), when 19 people were injured, and the tornadoes that hit the south coast of England in November 2006.

Floods

Periodic **floods** are common in some parts of the world. According to the UN, floods affect up to 500 million people per year worldwide — more than any other natural hazard. They result in as many as 25 000 deaths annually, extensive homelessness and disaster-induced disease.

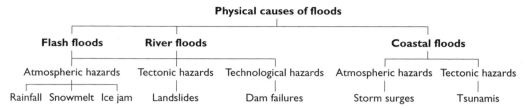

Figure 23 Physical causes of floods

Globally, the greatest potential flood hazard is in Asia, where, between 1995 and 2004, approximately 35% of all floods were reported, compared with just over 20% in Africa and the Americas. In China, the great floods of 1887, 1931 and 1959 each killed over 1 million people. Projects such as the Three Gorges Dam not only offer a sustainable power supply but also are important flood control measures.

Devastating floods also occurred in Mozambique (2000), Bangladesh (2004) and Bolivia (2006).

In 2005, there were floods in Carlisle, UK:
- The worst floods since 1822 began on 8 January 2005 — a 1-in-170-year event.
- Rainfall was 2800 mm on the 2400 km² catchment.
- The antecedent conditions were 2 weeks of heavy rainfall.
- Gales caused trees to fall. Surface runoff and erosion blocked streams and drains.
- River and groundwater levels broke existing flood defences.
- The Rivers Caldew and Peteril have their confluences with the River Eden in Carlisle. This worsened the impact.
- The damage affected 1900 properties; 2500 residents were displaced and three people died.
- Total losses were approximately £400 million.

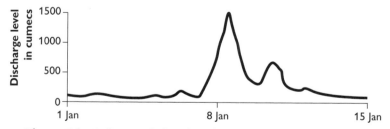

Figure 24 Hydrograph for the River Eden, Carlisle, 2005

Features of the hydrograph for the River Eden (Figure 24) include:
- a steep rising limb and recessionary limb characteristic of a 'flashy' response. This indicates high flood potential.
- a dramatic increase in discharge from an average of 100 cumecs (before 8 January) to a peak flow of 1500 cumecs. Thus, channel flow greatly exceeded channel capacity, so flooding was an inevitable consequence.
- a 7-day period for the river to return to normal flow levels.

Drought

Drought, like many natural hazards, is not purely a physical phenomenon; it is an interplay between natural water availability and human demands for water supply. It can occur virtually anywhere on the planet. There are different types:

- **Meteorological drought** occurs when there is a prolonged period with little or no precipitation associated with blocking anticyclones.
- **Agricultural drought** occurs when there is insufficient moisture for average crop or livestock production. This can arise even in times of average precipitation because of soil conditions or agricultural techniques.
- **Hydrologic drought** occurs when the water reserves available in sources such as aquifers, lakes and reservoirs fall below the statistical average. This can arise even in times of average (or above average) precipitation — for example, when increased water usage diminishes the reserves.

The word 'drought' is used mostly to mean meteorological drought. However, when the word is used by urban planners, it is more frequently in the sense of hydrologic drought.

Persistent drought can result in the slow starvation of whole communities. In early 2006, the UN warned that 11 million people were likely to die of hunger in the Horn of Africa (Figure 25) following a long drought. The countries particularly affected were Somalia, Kenya, Djibouti and Ethiopia.

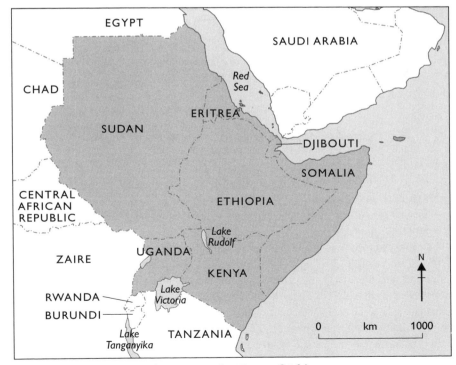

Figure 25 The Horn of Africa

To make matters worse, the rains that end a long drought may trigger a plague of locusts, further lowering agricultural output. In areas such as the Horn of Africa, drought is commonplace since there is a belt of high pressure near to the Tropic of Cancer.

In England in 2006, there was a significant threat of drought (Figure 26). The area at highest risk was the far southeast, particularly Kent. A primary influence was that in southeast England the preceding 18 months had been unusually dry. Reservoirs were only partially full, but it was the low levels of groundwater and rivers that caused most concern. In the southeast, much of the water supply comes from such sources.

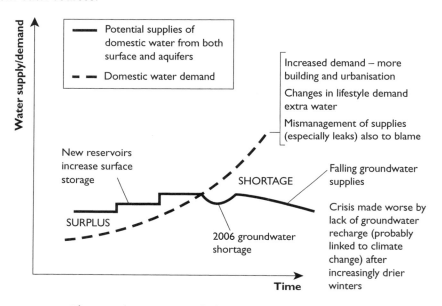

Figure 26 Water supply in the southeast of England

The 2006 drought was not as significant as that of 1975/76 for various reasons:
- In 2006, drought orders were issued earlier by water companies in an attempt to reduce consumption.
- August 2006 rainfall, particularly in the east of England, led to the partial recharge of depleted aquifers.
- The 1976 drought was made worse by the blocking anticyclone. The really big heat started on 23 June. During the next fortnight, temperatures reached 32°C across much of southern England. Then, it remained hot almost until September. August 2006 was relatively cool, dominated by showery weather and cooler northerly winds.
- A significant difference between the 2006 and 1976 events was the spatial extent. The 1976 drought impacted on much of England ('hosepipe patrols' operated in Birmingham, some northern cities had standpipes and there were numerous forest

fires), whereas in 2006 only the southeast was threatened. In 2006, much of the country had ample water, particularly surface storage (rainfall in May 2006 was exceptionally high — York was flooded), whereas in 1976 there was significant reduction in surface storage in most parts of the country.

There may be a direct link between global warming and the increased incidence of drought and desertification. Some of the most significant impacts, in areas already prone to aridity, will be an increased incidence of famine, increased infection and human displacement. This in turn may have secondary impacts, such as rural-to-urban migration, increasing urbanisation, reduction in tax revenues and unemployment.

Landslide and avalanche hazards

Landslides

Landslides are the seventh biggest killer ranking below almost all other natural hazards. Only volcano and drought hazards have lower global death rates.

However, some parts of the world suffer significant losses by landslide. A 35-year study of Japan (1967–2002) showed that, during that time, landslides occurred every year, killing almost 3300 people. Landslides also threaten some of the world's most precious cultural sites, including Egypt's Valley of the Kings and the Inca fortress of Machu Pichu in Peru.

On 17 February 2006, a landslide struck the village of Guinsaugon on the island of Leyte in the Philippines. A 9 km^2 ridge above the village collapsed after days of heavy rain and covered the settlement with mud and rock up to 30 m deep. An estimated 1000 people died. The incident was made worse because the area had suffered massive deforestation. There is also a rapidly growing population in the Philippines, which puts additional pressure on the resources of an already crowded island. The typhoons that hit the islands in 2006 caused similar problems. The secondary hazards of typhoon Durian (mud and debris slides) killed 400 people in December 2006.

The distribution of landslides in the city of La Paz, Bolivia is controlled by variations in slope gradient, the nature of overlying surface deposits and drainage-density patterns (Figure 27). Landslides are triggered by saturation of overlying deposits — sources include rainfall, stream water, and seepage from high, surrounding water tables and domestic sources. The most vulnerable people exposed to landslide hazards are the inhabitants of self-built informal housing areas who occupy the more elevated, steeper slopes of the northern part of the city. The southern part of the city is affected by slower forms of land failure. Rich housing in downstream areas is affected by flash flooding exacerbated by deforestation on the northern slopes.

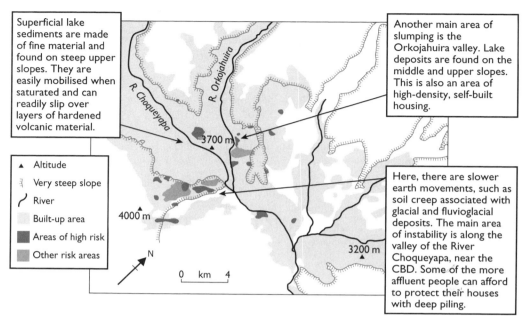

Superficial lake sediments are made of fine material and found on steep upper slopes. They are easily mobilised when saturated and can readily slip over layers of hardened volcanic material.

Another main area of slumping is the Orkojahuira valley. Lake deposits are found on the middle and upper slopes. This is also an area of high-density, self-built housing.

Here, there are slower earth movements, such as soil creep associated with glacial and fluvioglacial deposits. The main area of instability is along the valley of the River Choqueyapa, near the CBD. Some of the more affluent people can afford to protect their houses with deep piling.

▲ Altitude
꠸ Very steep slope
/ River
Built-up area
Areas of high risk
Other risk areas

Figure 27 Distribution of landslide areas in La Paz

Avalanches

Avalanches are not hazards until they affect people in some way:

- Avalanches happen quickly with little warning.
- They are limited in spatial extent and impact.
- There is a huge release of energy in a short space of time.
- The chances of survival are limited for anyone caught in an avalanche.
- Avalanche prediction is reasonably accurate over large areas.
- Mitigation responses cover 'hard' and 'soft' engineering, planning, warning and emergency action at several levels.

The spatial distribution of avalanches is strongly concentrated in mountainous areas since they are associated with the failure of snow slopes. Avalanches almost always occur on slopes of 30–40°, because little snow clings to steeper slopes and the forces needed to move snow on lower-angle slopes is much greater than is normally achieved. They occur when the stress on the snowpack exceeds the resistance holding the snow on the slope. Causes of avalanche can be small, perhaps a skier or snowboarder just tipping the balance of forces.

Major avalanche events occur somewhere in Europe or North America nearly every year and cause a number of deaths (on average 40 in Europe and 100 in North America). In February 1999, more than 30 people were killed in Galtur, Austria. This avalanche moved at speeds of up to 200 m s⁻¹ and involved almost 0.75 million tonnes of snow. The event was particularly significant because it destroyed part of a village that was thought to be safe from avalanches. The weather, type of snow and speed of the avalanches all contributed to the disaster, but the major problem was the 'certainty' that no avalanche could reach the village. As a result of this incident, avalanche planning in Europe was revised.

Generalisation 2: Spatial variations in the impact of natural hazards

Survey of physical factors influencing impacts

There are three main parts to this generalisation:

- Impacts (including number of deaths, economic damage) are related to the level of development of an area, as well as to its physical nature (e.g. location/ characteristics) — see Figures 28 and 29.
- The impacts of natural hazards can be divided into those that are social, economic and environmental consequences.
- Spatial variations include the location, type, frequency and magnitude (size) of the natural hazard.

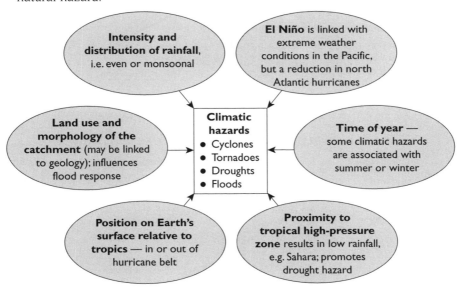

Figure 28 Factors that affect climatic hazards

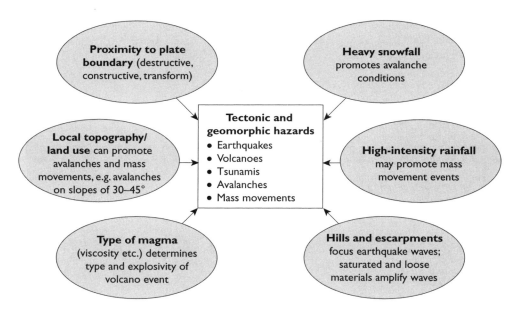

Figure 29 Factors that affect geomorphic hazards

Spatial variability between hazard events

All hazards have the potential to kill people and to cause damage and economic impact. Yet some similar hazard events may have very different impacts. Both physical and human geography play a part in the severity of impact. For example, the removal of coastal mangroves can increase the impact of coastal flooding from storm surges (e.g. in Florida). The distribution of population also plays a major part in determining the potential hazard impact and associated economic loss.

Hurricanes

Table 2 shows the variability in impacts of a series of hurricanes in the 2004 season.

Table 2 The 2004 hurricane season in the USA and Caribbean

Storm	Dates* (2004)	Location	Intensity	Death and damage	Key features
Charley ('the capricious')	12–14 August	Predicated to hit Tampa; came in at Punto Gorda and then moved up the east coast	Downgraded from category 5 to 4	27 killed in Florida; up to $25 billion in lost tourism	Intense coastal damage; unpredictable

Storm	Dates* (2004)	Location	Intensity	Death and damage	Key features
Frances ('wide-span Fran')	4–6 September	Landfall at Stuart (east coast), then moved to west coast	Downgraded from category 4 to 3	Four deaths; widespread flooding; damage £2–4 billion	Good preparation: 2.4 million evacuated; 4–5 million lost power
Ivan ('the terrible')	7–16 September	Came via Grenada, Jamaica, the Cayman Islands to the Florida panhandle and looped the loop	Category 5	Overall, 122 dead including 34 in Grenada and 49 in the USA	Lots of warning, but wreaked havoc in the Caribbean Islands
Jeanne ('the slow, wet one')	17–26 September	Slow through Haiti (17–19), then to the Bahamas and the east coast of Florida	Tropical storm, except in Haiti where category 3	3000+ deaths, including 1800 in Haiti (largely from flooding and landslides)	Difficult to predict, because slow; 3 million evacuated from Miami

* Dates of the damaging storm event, not the duration of the whole tropical depression

Key points
- The strongest storms do not always cause the greatest damage and the most deaths. Jeanne was mostly category 2 (except over Haiti), but it led to the greatest number of deaths. The size of the cell, the speed of movement, and the amount and intensity of rainfall are crucial.
- Sequencing is important. Much of Florida had already suffered multiple events and evacuation fatigue had set in. This was compounded by unreliable forecasting.
- In spite of the modern hurricane prediction facilities in Miami, hurricane tracks and speed remain difficult to predict.

Variations in earthquake impact

Earthquakes are the most frequent of all natural hazard events. Their impacts on people, property and communities vary enormously.

Physical factors
- **Plate tectonics**   Distribution is linked to destructive or transform plate margins. However, some smaller earthquakes (e.g. Dudley, 2002) have occurred away from margins along ancient faults.
- **Earthquake magnitude and depth**   In general, stronger earthquakes have more serious effects. However, shallow earthquakes close to the surface cause more intense shaking and more damage.

- **Nature of bedrock**   Some materials (e.g. clay and silt) are prone to liquefaction. Many buildings in the 1985 Mexico City earthquake became tilted on the lake-bed sediment.

Human factors

- **Population density**   There is considerable overlap between major earthquake zones and areas of high population density. Seventy of the world's largest cities (with 10% of the population) are in earthquake zones.
- **Remoteness**   When earthquakes occur in remote areas (e.g. Afghanistan, 2002 and Kashmir, 2005), it is difficult for rescue teams to get help to people, because roads and other infrastructure have been damaged by the earthquake.
- **Building and structural vulnerability**   Modern buildings with appropriate design can minimise loss of life (e.g. Loma Prieta, California, 1989, when only 67 died). In LEDCs, design is often inadequate, with regulations only weakly enforced. In Mexico City in 1985, several modern, high-rise buildings collapsed and 30 000 lives were lost.
- **Extent of earthquake preparedness**   In wealthy areas where earthquakes are common (e.g. California and Japan) there are regular drills and people are well informed. Poorer countries tend to be less well prepared, mainly because of a lack of investment.
- **Levels of development**   Generally, LEDCs suffer the impacts of earthquakes much more than MEDCs. This is certainly true for more recent events (see Tables 3 and 4). However, MEDCs suffer massive financial losses as insurance companies and governments fund rebuilding programmes and pay compensation.

Data from the last 50 years do *not* show that the frequency of earthquakes is increasing. Human societies are becoming more vulnerable, rather than there being increased seismicity. This escalation of vulnerability is linked to human factors — population growth, urban growth, overcrowding of cities and the increasing sensitivity of modern, technologically based societies.

Table 3   The largest earthquakes, 1990–2005

Year	Magnitude	Fatalities	Region
2005	8.6	1 313	Northern Sumatra
2004	9.1	283 106	Near west coast of northern Sumatra
2003	8.3	0	Hokkaido, Japan
2002	7.9	0	Central Alaska
2001	8.4	138	Near coast of Peru
2000	8.0	2	Papua New Guinea
1999	7.6	2 297	Taiwan
1999	7.6	17 118	Turkey

Year	Magnitude	Fatalities	Region
1998	8.1	0	Balleny Islands
1997	7.8	0	South of Fiji
1997	7.8	0	Near east coast of Kamchatka
1996	8.2	166	Irian Jaya
1995	8.0	3	Near coast of northern Chile
1995	8.0	49	Near coast of Mexico
1994	8.3	11	Kuril Islands
1993	7.8	0	South of Mariana Islands
1992	7.8	2519	Flores, Indonesia
1991	7.6	75	Costa Rica
1991	7.6	0	Kuril Islands
1990	7.7	1621	Luzon, Philippines

Table 4 The deadliest earthquakes, 1990–2005

Year	Magnitude	Fatalities	Region
2005	7.6	80361	Pakistan
2004	9.1	283106	Near west coast of northern Sumatra
2003	6.6	31000	Southeast Iran
2002	6.1	1000	Hindu Kush, Afghanistan
2001	7.6	20023	India
2000	7.9	103	Southern Sumatra
1999	7.6	17118	Turkey
1998	6.6	4000	Afghanistan–Tajikistan border
1997	7.3	1572	Northern Iran
1996	6.6	322	Yunnan, China
1995	6.9	5530	Kobe, Japan
1994	6.8	795	Colombia
1993	6.2	9748	India
1992	7.8	2519	Flores, Indonesia
1991	6.8	2000	Northern India
1990	7.4	50000	Iran

Why are some volcanoes more hazardous than others?

Some volcanoes merely 'fizz' while others go 'bang'. Why is this? The nature of the volcanic eruption is determined by the tectonic context (Table 5; Figure 18).

Table 5 A comparison of volcanoes

Example	Magma type	Tectonic context	Notes
Kilauea, Hawaii	Basaltic	Oceanic hot spot	• Hazard levels are low, with the basalt erupting almost continually • The lava can flow for several kilometres overwhelming homes, roads and forest • Deaths are rare
Nyiragongo, Democratic Republic of Congo (DRC)	Basaltic	Continental constructive plate margin	• Very mobile basalt can travel up to 60 km h^{-1} • This volcano erupts every 15–20 years as the continental crust is pulled apart and becomes thin • Rapidly growing population is a problem
Soufrière, Montserrat	Andesitic	Island-arc destructive boundary	• Montserrat eruptions (1995–97) were a major tragedy • The island and its capital (Plymouth) were overwhelmed by pyroclastic flows • Andesitic magma is thick and sticky • Subduction causes wet partial melting, converting basaltic oceanic plate into high-gas and high-silica content (andesitic) magma
Yellowstone, USA	Rhyolitic	Continental hot spot	• Yellowstone is a 'supervolcano' • The hot spot beneath the continental plate melts the plate above • This generates rhyolitic magma • As it slowly rises, its causes the surface to bulge, eventually failing • The result is rapid gas decomposition and an eruption of great violence

Table 6 The effect of magma type on volcanic eruption

Magma type	Eruption temperature (°C)	Gas content	Viscosity	SiO_2 content (%)
Basaltic	1100	Low	Low/runny	50
Andesitic	1000	Moderate	Moderate	60
Rhyolitic	800	High	High/sticky	70

Magma type determines the volcanic explosivity index (VEI) (Figure 30). Explosivity is related to magma viscosity. Magmas with a high gas content, high silica content and low temperature are viscous and produce explosive, hazardous volcanoes.

Impact and loss of life from volcanic incidents are much less significant than from other hazards. Data about volcanic incidents from 1992–97 include:
- average number of people killed per year — 1019
- average number of people injured per year — 285
- average number of people affected per year — 94119

The more viscous the magma (e.g. rhyolite), the rarer the eruption (Figure 31).

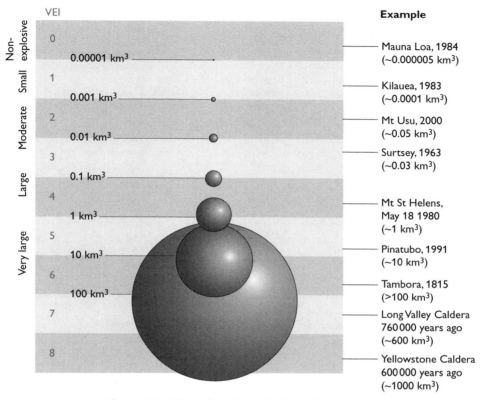

Figure 30 The volcanic explosivity index (VEI)

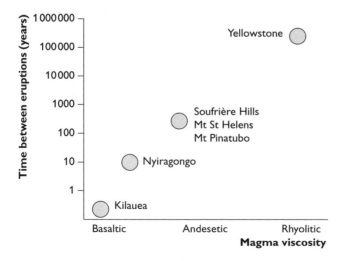

*Figure 31 The relationship between magma viscosity
and time between eruptions*

Table 7 shows how volcanic hazards can have a variety of spatial impacts, ranging from local to global.

Table 7 Volcanic hazards: potential impacts and scale

Hazard type/ category	Local impact	Disaster potential (region/country)	Catastrophe potential (cross-boundary)	Global consequence
Gas discharge	✔	✔		
Tsunami	✔	✔	✔	
Lahar	✔	✔		
Flooding	✔	✔		
Pyroclastic flow	✔	✔		
Landslide	✔	✔		
Earthquake	✔	✔		
Lava flow	✔	✔		
Ashfall	✔	✔	✔	
Explosivity	✔	✔	✔	✔

Spatial variability of impact: the Asian tsunami 'mega' event, 2004

The impact of tsunamis is limited geographically, affecting only landmasses at the edge of some of the oceans. However, where they do strike, it can be with a destructive force greater than that of any other type of disaster.

The Asian tsunami of 26 December 2004 sent waves ricocheting around the world in an unusual pattern — there was not a straightforward decay of wave height with distance from the tsunami epicentre (Figure 32). An explanation for this is that underwater structures such as mid-ocean ridges and continental shelves funnelled the waves across massive distances. It is thought that there are two main factors that affect the reach of tsunami waves:

- the direct impact of the force or size of the earthquake
- the topography of the sea floor and how it guides waves

Wave heights of about 50 cm were recorded in the Cocos Islands (1700 km from the epicentre); in India and Sri Lanka (a similar distance from the epicentre), waves of up to 5 m were recorded.

The information presented in Table 8 shows how the tsunami affected the people, the environment and the economies of countries at different levels of development. Along the crowded coastlines, the impact on the people and their economies was disproportionate.

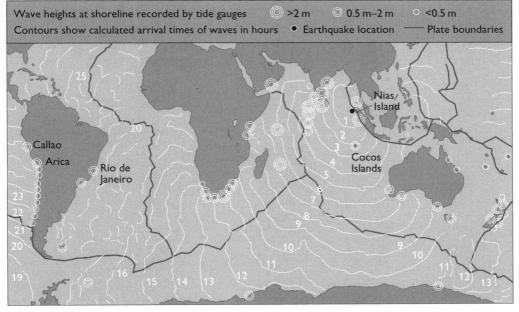

Figure 32 How the tsunami spread round the world

Table 8 The human and economic impact of the tsunami at different locations

	Indonesia	Burma	India	Maldives	Sri Lanka	Thailand
Deaths	169000	81	10750	81	31000	5300 (including 2248 foreign nationals)
Displaced persons	600000	10000 –15000	140000	11500	500000	300000
Homes damaged/ destroyed	100000+	5000	15000	15000	100000	60000+
Employed in tourism/ travel (%)	Minimal	3.1	5.5	54.2	7.8	8.4%
Notes	Aceh province was a poor, war-torn area	Only limited information available	India is an NIC, but the area affected was poor	Tourism is important to the local economy	Widespread damage around the coast	International tourists died here

Understanding multiple-hazard zones

Traditionally, multiple-hazard zones (MHZs) are regions that are exposed to a range of hazards (often a combination of meteorological, climatic and geomorphic impacts).

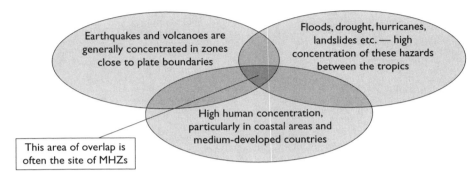

Figure 33 The site of multiple-hazard zones

Why are some areas more hazardous than others?

Most MHZs result from an interaction of climatic and geomorphical/geophysical hazards. Areas where there is a high population density will suffer most from multiple hazards. This effect is compounded within 'medium' developed countries (typically RICs and NICs) because they often have megacities and limited or inadequate disaster-prevention infrastructure. Figure 34 shows the top six multihazard 'hotspots' (Table 9).

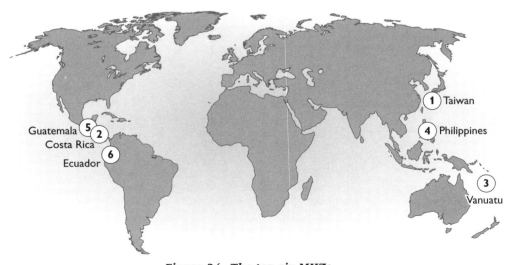

Figure 34 The top six MHZs

Table 9

Country	Percentage of area exposed	Percentage of population exposed	Maximum number of hazards (types)
Taiwan	73.1	73.1	4
Costa Rica	36.8	41.1	4
Vanuatu	28.8	20.5	3
Philippines	22.3	36.4	5
Guatemala	21.3	40.8	5
Ecuador	13.9	23.9	5

Case study: the Philippines, a 'conventional' MHZ

The Philippines in southwest Asia consists of over 7000 small islands in a band 5–20°N of the equator that is within the belt of tropical storms. The country is lower–middle income with a high population density (240 km^{-2}) and has a rapid population increase. It is located on a destructive plate boundary.

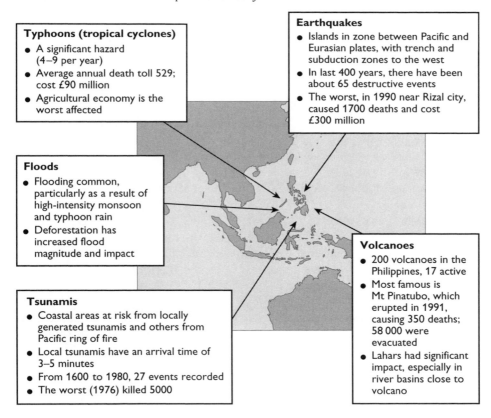

Typhoons (tropical cyclones)
- A significant hazard (4–9 per year)
- Average annual death toll 529; cost £90 million
- Agricultural economy is the worst affected

Earthquakes
- Islands in zone between Pacific and Eurasian plates, with trench and subduction zones to the west
- In last 400 years, there have been about 65 destructive events
- The worst, in 1990 near Rizal city, caused 1700 deaths and cost £300 million

Floods
- Flooding common, particularly as a result of high-intensity monsoon and typhoon rain
- Deforestation has increased flood magnitude and impact

Volcanoes
- 200 volcanoes in the Philippines, 17 active
- Most famous is Mt Pinatubo, which erupted in 1991, causing 350 deaths; 58 000 were evacuated
- Lahars had significant impact, especially in river basins close to volcano

Tsunamis
- Coastal areas at risk from locally generated tsunamis and others from Pacific ring of fire
- Local tsunamis have an arrival time of 3–5 minutes
- From 1600 to 1980, 27 events recorded
- The worst (1976) killed 5000

Figure 35 The Philippines: an MHZ

Other types of MHZ

Another interpretation of MHZs is that some parts of the world are exposed to the *same* hazard repeatedly over a relatively short space of time. For example, in 2004, Florida took three major hits during the hurricane season.

Multiple hazards in the UK

Some countries that are usually considered to be non-hazardous may have 'micro-hazard hotspot zones'. An example of one such area is West Sussex, UK. Between 1963 and 2000 this area suffered two earthquakes, a storm, two river floods and four tornadoes.

The immediate causes of these events can be explained easily, so perhaps it was just chance. Another view is that an impact of global warming is the generation of more extreme weather events, making them a more common occurrence.

What role did humans play in increasing the risk of damage from these hazard events? There is no doubt that densely populated areas can exacerbate flood risk, because building on the floodplain leads to increasingly rapid runoff.

Some researchers have tried to link the increased instability of weather in the UK to El Niño. The large number of tornado incidences between 1998 and 2000 could be linked to this. However, it may also be related to increased public awareness, with more people reporting incidents to the authorities.

Creating a multiple-hazard diary: Japan, 2004

The following list is an example of a multiple-hazard diary for hazard events in Japan in 2004 (see Figure 36).

- 19 July: **flooding** — 18 dead; 490 mm rain (about 20% of total annual rainfall) fell in 4 days in Niigata state
- 19–20 August: **Typhoon Megi** — 90 mph winds; 205 mm rain; localised landslides and flooding
- 1 September: **volcanic eruption** — minor eruption of Mt Asama; no evacuation necessary; 183 minor earthquakes associated with event
- 1–3 September: **Typhoon Chaba** — nine dead; 204 injured; 19 000 homes flooded; winds up to 130 mph damaged property and crops
- 6 and 8 September: **earthquakes** — 6.9 and 7.3 scale earthquakes hit southwest offshore Japan; depth and distance from coast reduced impacts; 1 m tsunami
- 7–9 September: **Typhoon Songda** — 225 injured; 83 000 households evacuated; 1.25 million people without electricity; wind speeds of 134 mph in southwest Japan

Tip

Creating your own multiple hazard map will help to raise your spatial awareness of the events.

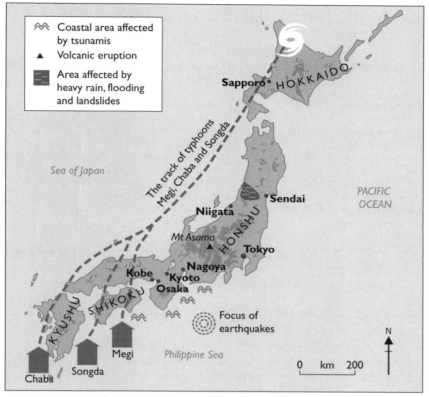

Figure 36 Hazard events in Japan, 2004

Generalisation 3: The human response to hazards

This generalisation contains a number of ideas which are summarised in Figure 37. In particular, you should be aware that:

- Responses are influenced by a complex and interlinked range of physical and human factors.
- Speed and effectiveness of response may also be a function of the different 'players' or groups/organisations involved in the immediate and longer-term management and recovery.
- The nature of the response will also be controlled by the degree of understanding of the hazard, its frequency, location (and accessibility/remoteness) and the economic environment.

Factors that influence the response

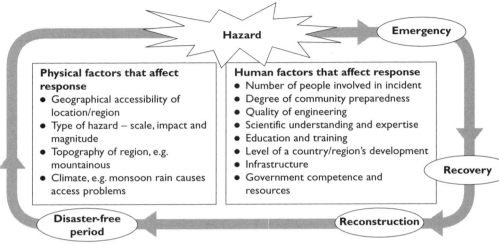

Figure 37 Factors that influence human response to hazards

The responses to hazards are critically controlled by the capability of the groups involved in their management:

- **National government** — national policies, civil protection and defence, and public information. The Federal Emergency Management Agency (FEMA) was highlighted with Katrina. In the UK, the Environment Agency and DEFRA have responsibility for flood warning and management.
- **Local government** — operational policies at the local level (e.g. rescue, welfare, medical, transport, and supply of emergency aid)
- **Scientists/academics/educators** — researching, understanding causes, producing hazard (risk) maps, dissemination of information to the public and raising awareness
- **Insurers** — risk assessment prior to hazard; finance and assistance after hazard
- **Planners** — reducing risk by land-use zonation and planning (e.g. restricting development on floodplains)
- **Relief agencies** — post-disaster aid (e.g. the International Red Cross featured during the 2004 tsunami)
- **Engineers and architects** — design of buildings and infrastructure; range from Japanese technologists to US Corps of Engineers (rebuilding of New Orleans flood defences)
- **Emergency practitioners and services** — police, medics, fire officers, traffic control and possibly the army; coordination is an important role
- **Media** — highlight the hazard event (magnitude, scale, location, impact, duration etc.); have a warning role that is important for more remote communities
- **Communities** — important role in managing situations and in terms of community preparedness and education

Case study: the earthquake in Kashmir, 2005 — a challenge?

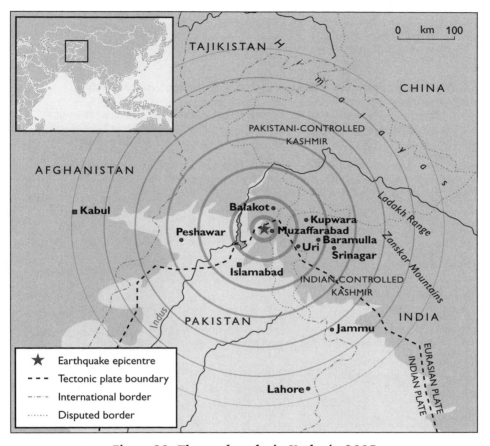

Figure 38 The earthquake in Kashmir, 2005

There was only limited success in response to the Kashmiri earthquake because:

- the relief agencies (headed by the UN) failed to raise sufficient aid ($550m needed) to cope with the disaster. The US was also criticised for responding with too little, too late. This increased the number of secondary deaths.
- in Pakistan, the army's initial response was slow
- the Pakistani government was reluctant to accept aid from countries such as Israel and India (owing to a perceived political risk)
- the 80 000 death toll was so high because schools and hospitals were shoddily built

The situation in Kashmir was particularly difficult to manage because:

- the area is politically unstable and is a war zone
- the area is isolated and has inhospitable terrain; landslides are frequent
- in some areas there is neither power nor adequate food and water, with a danger of disease spreading

- the incident coincided with the start of winter and the onset of extremely cold weather, exacerbated by the high altitudes
- the numerous aftershocks hampered rescue efforts
- the magnitude of this disaster was so vast that the governments alone (especially Pakistan and India) could not be expected to provide adequate relief

Case study: the Boscastle flood, 2004 — a success?

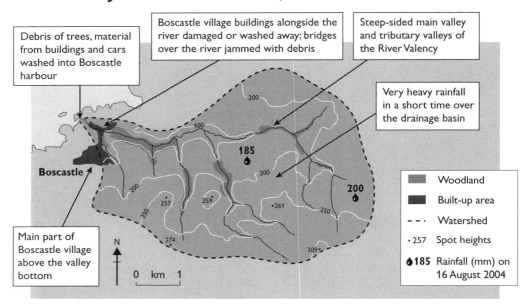

Figure 39 The flash flood in Boscastle, 2004

The response to the Boscastle flood was successful. The timeline of the events on 16 August 2004 was as follows:

- 2.00 p.m. — two Environment Agency (EA) operatives sent to check flooding of drains
- 3.15 p.m. — EA opens a flood incident room in Bodmin
- 3.44 p.m. — local coastguard warns Falmouth coastguard of an incident developing in Boscastle
- 3.53 p.m. — five local firecrews sent to Boscastle
- 4.00 p.m. — visitor centre manager ushers families into the attic to escape the floods
- 4.45 p.m. — first of seven helicopters from the Royal Navy, RAF and Coastguard arrives
- 5.12 p.m. — fire and coastguard services declare a major incident and inform the media
- 5.23 p.m. — rescue helicopters begin winching people up from buildings
- 5.55 p.m. — Truro and Plymouth hospitals are put on standby in case of casualties

The fact that there were no fatalities from this incident is largely due to the 'textbook' coordination of rescue and monitoring efforts from a range of agencies. A contributory factor may also have been the geographical situation of Boscastle — close to an RAF station with service personnel trained in rescue.

Other case studies

Two further case studies that deserve particular mention are the 2004 tsunami and Hurricane Katrina:

- The tsunami was a mega-event. In the immediate aftermath, confusion arose over who would coordinate relief operations. The UN's Office for the Coordination of Humanitarian Affairs (OCHA) was criticised for not having the resources or authority to deal immediately with such an event. (For details, see pp. 61–63.)
- Katrina exposed the shortcomings between local and state government. The mayor of New Orleans was critical of the coordination of relief efforts. The speed of relief was inadequate, causing unnecessary death and suffering. Katrina also cast doubt on the USA's ability to cope well in an extreme situation. (For details, see pp. 63–64.)

The response management framework applied to two mega-events

Case study: the Asian tsunami, 2004

On 26 December 2004, the most powerful earthquake for over 40 years (over 9.0 on the Richter scale) struck off the coast of Sumatra. Figure 40 shows the extent of the impact; over 275 000 people were killed. The event had other short-term and long-term impacts on the environment, tourism, education, health, agriculture, housing, social infrastructure, power, water supply, transport (road and rail) and fisheries.

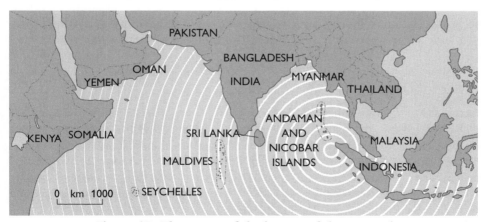

Figure 40 The extent of the impact of the tsunami

Figure 41 shows the disaster-response curve (Park response model). This is an attempt to model the impact of a disaster from pre- to post-disaster. It also considers the role of emergency relief and rehabilitation. For each hazard event, or in the case of a multi-regional event, each area or country may have a different response curve.

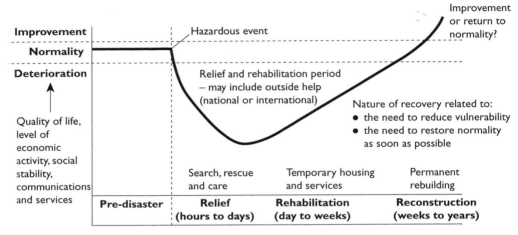

Figure 41 A disaster-response curve

Stage 1: pre-disaster

Key factors that influenced the severity of the impact included:
- physical geography of the area — low-lying coastal plains
- low levels of development of most countries hit by the tsunami
- high population densities — up to 900 people per km^2 in Sri Lanka
- some areas of coast (e.g. Sri Lanka and Aceh) were war-torn and distressed

Stage 2: impact of the disaster

Physical factors were important here — for example, distance from the epicentre. Aceh was hit both by a mega-earthquake and by the tsunami. Further from the epicentre there was more time for communication networks to deliver warnings, although these were not always given (e.g. Somalia). In Thailand, there were additional complications with impacts at Phuket (an international holiday resort) on a national holiday. However, this had a positive impact by raising the awareness of the event and undoubtedly increasing the amount of international aid.

Stage 3: emergency relief

Oxfam stated that the emergency relief effort was successful, with few secondary deaths from starvation and disease. However, the geographics of recovery were varied. There was a differentiation in terms of response rates and the effectiveness of programmes. The cohesiveness of the community and the access to social, economic and political resources were also crucial. Where infrastructure and communications were damaged (e.g. Phi Phi, Thailand), NGOs and government organisations lacked planes and boats to reach the communities.

Stage 4: rehabilitation (after 1 year)

This phase illustrated the differential progress among communities within the same country and among different countries. The Cash for Work programme was a major success — survivors were found work building boats, desalinating land, building and organising village communities as trade cooperatives. In some areas, 60–70% of fishermen were back in business, some recording catches of 70% of pre-tsunami levels. However, there is a deficit in funding for some projects in Aceh and Nias. According to the World Bank, the biggest shortfalls are in transport and flood control.

Stage 5: recovery

In terms of mitigation, enormous progress has been made with the provision of a tsunami warning system. There are hopeful developments for improving the effectiveness of the Pacific Tsunami Warning System (based in Hawaii) and for the development of a new Indian Ocean Warning System, spearheaded by the UN.

Case study: applying the Park model to Katrina

An ironic factor in the New Orleans case is that the event had been foreseen both in the short and long term. In the long term:

> New Orleans is a disaster waiting to happen. The city lies below sea level in a bowl bordered by levels that fen off Lake Pontchartrain in the north and the Mississippi to the south and west. The city is sinking...putting it at risk from even minor storms.
>
> *Drowning New Orleans*, 2001

In the short term, the hurricane-warning centre predicted a big hit close to the city days before the event.

On 28 August 2005, Mayor Nagin ordered a mandatory evacuation of the city.

Impact of the disaster

The storm hit on 29 August 2005, with wind speeds of up to 145 mph (category 4). However, the real impact was not the wind but the rising floodwaters (both inland and coastal). Biloxi recorded the largest ever storm surge (10 m) in the USA. On 30 August, a major levée failed and 80% of the city was flooded. An estimated 600 000 fled the immediate area of the city.

The official death toll in Louisiana, Mississippi and Alabama was just over 1000. There were also significant economic consequences. Immediately after the event, the price of oil jumped to a record $70 per barrel. The estimated cost of the disaster is believed to be about $25 billion.

Emergency relief

The American Red Cross quickly set up 275 shelters in nine states and supplied 249 emergency vehicles, 4200 relief workers and 140 000 meals.

On 31 August, government help started to arrive (organised by FEMA), amid much criticism — there were too many people on the Gulf coast when Katrina struck and too many were poor or immobile. The government did not provide enough help to get them out. The breakdown of law and order in the city was particularly disturbing. Armed police were ordered to tackle the lawlessness. Military transport planes evacuated only the seriously sick and injured to Houston. On 2 and 3 September, the military took over the city. Refugees were removed to other cities.

Rehabilitation and recovery

Katrina has highlighted the racial divide that still exists in the USA. Poor black people were in the most trouble; the wealthier white population had a better chance. By March 2006, less than half the city's former 450 000 residents had returned.

However, there has been a range of rehabilitation and rebuild events:

- February 2006 — Mardi Gras took place. It was needed to kickstart the local economy. New Orleans was once more open for business.
- March 2006 — FEMA released revised flood-risk maps. These have an impact on new building codes and standards as well as defining permissible locations.
- April–May 2006 — the Bring New Orleans Back Commission plan was launched. This is to help oversee the city's redevelopment and plays a role in deciding which part of the city is to be revived first.
- June 2006 — funds from the western Gulf oil and gas lease will help pay for restoration of Louisiana's marshes and barrier islands, which act as natural buffers. This was the deadline for completion of levée system repairs by the US army corps.
- September 2007 is the deadline for returning all the levées and floodwalls to their original design height, although in the future, defences will have to be raised further because the city is subsiding.

In the longer term, coastal restoration will create a buffer against storms and will protect rural areas, ring-fencing the city with stronger levées.

Assessment of the role of community preparedness and education

Community preparedness

Disaster reduction is most effective at the community level, where specific local needs can be met. When used alone, government and international agencies often prove to be inadequate. They may be inclined to ignore local perceptions and needs and the potential value of local resources (skills and knowledge). As a result, it is not surprising that emergency-relief assistance far exceeds the resources invested to develop local capabilities to reduce disaster risk.

Mini case study: Norway avalanches

Snow and slush avalanches are a natural hazard to local communities in parts of Norway. They cause human fatalities and significant damage to houses and infrastructure every winter.

Geiranger is an area on the west coast of Norway with a high exposure to snow avalanches. Relocating 1000 residents is not realistic, so energy was devoted to finding acceptable means by which they could live, with minimised risks. The assessment concluded that building mitigation measures could not be justified because of the high cost set against the low frequency of events. Instead, an early warning system together with a preparedness plan based on community actions was adopted:

- Get technical help to make a hazard map for avalanche-prone areas.
- Organise a local avalanche group consisting of representatives from the community, including the police and civil defence agency.
- Install meteorological equipment to help with avalanche prediction.
- Develop an action plan for different hazard levels, including procedures for warning, evacuation and training of the local avalanche group.

The system was put to a real test on 4 March 2000. The hazard level was high and 32 people were evacuated to a hotel in a safe area. An additional 180 people were trapped between two avalanches because of an impassable road, but were successfully evacuated. Because of the well-developed preparedness plan, all operations were successfully carried out without the loss of life.

The learning experience from this case is positive. Several other communities along the western coast of Norway are now adopting a similar approach.

Mini case study: Mozambique

The response to the floods in 2000 in Mozambique — the worst for a century — was in some ways a success. Of great significance were the 45000 lives saved by rescue efforts coordinated and delivered by regional rather than international rescuers. In 2001, another wave of devastating floods hit a different part of Mozambique. Local teams, operating mainly by boat, rescued over 7000 people. In each year, for every person who died, over 60 were saved. While media images of helicopters rescuing poor Africans gave the impression that international aid agencies saved the day, the real story is very different. Despite being one of the world's poorest countries, Mozambique was well prepared. Although international help was crucial, it succeeded because agencies let Mozambicans take the lead.

The success was due to a number of central factors, including:

- having safe places for people and cattle during an evacuation
- clearly marked escape routes
- legal powers that force people to move

Success was also attributed to simulations of major flood incidents involving the police, Mozambique Red Cross, local flying clubs, the fire brigade and Scouts.

There has been criticism of the community preparedness scheme:

- Mozambique has to choose between disaster preparedness and other initiatives, such as health care and sanitation.
- There is an acute lack of financial resources — flood preparedness has dropped down the list of priorities.

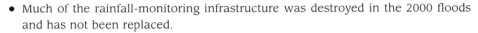

- Much of the rainfall-monitoring infrastructure was destroyed in the 2000 floods and has not been replaced.

Mini case study: Gujarat, India

In January 2001, immediately after the Bhuj earthquake in Gujarat, Indian community-based organisations began to help in the recovery effort. A policy was proposed that would not only rebuild the devastated Gujarat communities, but also reform and strengthen their social and political structures. The central concept was that people need to rebuild their own communities. Key elements of the strategy included:

- using reconstruction as an opportunity to develop local capacities and skills
- forming village development committees, made up of women's groups, to manage rehabilitation
- engaging village committees in monitoring earthquake-resistant construction
- striving to locate financial and technical assistance within easy reach of affected communities and not being dependent on it being mediated by others
- distributing information about earthquake safety
- encouraging the use of local skills and labour, and retraining local artisans in earthquake-resistant technology
- encouraging coordination among government officials, district authorities and NGOs and seeking to facilitate public–private partnerships for economic and infrastructure development

This example shows how community involvement has added value — not just by contributing to the rebuilding and rehabilitation of the communities, but also in strengthening relationships between local stakeholders and the government and in empowering and engaging women.

Benefits and limitations of community preparedness programmes

Benefits include:

- the integration of community-based disaster preparedness with health programmes promotes development and income generation, increasing resilience to disasters
- bridging the gap between relief and rehabilitation
- participatory rapid appraisals, which provide relevance, increase capacity building and motivate self-initiated projects
- action planning and identification of vulnerability become more problem-oriented and focused
- the establishment of better links and networks with local government leaders

Limitations include:

- some misunderstandings with local authorities, who see the programmes as a threat to maintaining a sense of dependency by the local population
- poor planning processes in some areas
- insufficient efforts/resources to ensure sustainability of projects after the initial funding period

- roles sometimes clashing with those of local authorities, especially in the absence of an inclusive planning process

Role of education in hazard management

The impacts of past events have revealed the important context of education for disaster-risk reduction: children who know how to react during an earthquake, community leaders who have learned how to warn their neighbours in a timely manner, and societies familiar with preparing themselves for natural hazards all demonstrate how education can make an important difference in protecting people at the time of a crisis.

Education for dealing with risk and disaster preparedness requires long-term invest-ment in both MEDCs and LEDCs. Cultural norms and values, as well as related risk perceptions, must shift — a process that cannot happen overnight. Education requires a constant and consistent approach, beginning at an early age and continuing through generations.

Mini case study: practical training in Nepal

Nepal is becoming more vulnerable to earthquake risk because of the increasing population, uncontrolled urban development and construction practices that have deteriorated over the last century. Despite this growing risk, until 1993, there was no institution concerned with this issue. Then, the National Society for Earthquake Technology (NSET) was established to confront the problem.

NSET has been involved in national and international projects, including one in Kathmandu that led to the development and implementation of the Kathmandu Valley Earthquake Risk Management Action Plan in 1998. More than 100 engineering students participated in a building inventory and vulnerability analysis programme. In particular, they were involved in aspects of safer construction in earthquake-prone areas.

A particular benefit of this programme has been the spread of information to local communities.

Mini case study: the UN Disaster Management Training Programme (DMTP)

The DMTP supports capacity-building efforts through international organisations and individual disaster-prone countries. Workshops have promoted the establishment of national or regional centres and strengthened their capacities to study technological and environmental hazards, seismic protection, crisis prevention and preparedness. DMTP has conducted more than 70 workshops involving 6000 participants in Africa, Latin America and the Caribbean, Asia and the Pacific, the Middle East and the Commonwealth of Independent States.

The DMTP provides professional and structured learning and skill-building programmes that also serve to:

- support and create synergies among partner organisations
- raise the profile and visibility of disaster management in areas of particular risk
- promote awareness-raising, motivation, adaptability, increased ownership and responsibility
- encourage the commitment of people, local and international resources, technologies and funding

The technological fix

Anticipatory risk management to reduce vulnerability relies on new technology, including geographical information systems (GIS), global positioning systems (GPS) and remote sensing (RS). Improved computer processing and storage provide opportunities to improve:
- project planning
- real-time decision making in emergencies (e.g. damage assessment and evacuation routes)

The increasing reliability of portable radio-based transmission systems and satellite phones allows communication even when ground-based infrastructure is inoperable or damaged. GPS technology is now much cheaper and widely available, as well as being portable. Its use with notebooks or handheld computers allows accurate position fixing and direction finding in remote areas.

Remote sensing in disaster reduction

Monitoring is essential for the detection of the onset of disasters. It also helps in appraising a disaster situation and in effective recovery and reconstruction. It is often necessary to monitor vast areas; airborne observation systems, including satellites, are the most effective for this.

Applications of remote sensing (RS) in disaster reduction and management include:
- **cyclone detection and warning**   Meteorological satellites play the main role in detecting and tracking cyclones as they form and move over the ocean.
- **weather radar**   A weather radar provides a quick estimate of rainfall density covering a large area. The continuous measurement of rainfall movement has been used to develop predictive models for short-term rainfall forecasts.
- **drought monitoring**   RS data cannot be used for early warning of drought, but they can be used to appraise the extent of drought impact. At national level, the main objective is to secure food supply, mostly through international assistance. Monitoring is achieved through interpretation of satellite images.
- **earthquake damage mitigation**   Satellites provide general information on land use that would support various civil engineering works for mitigation and response. Interferometry synthetic aperture radar applications have been used to view small (centimetre level) crust movements that have occurred during earthquakes. This technology has so far had limited use in earthquake prediction.

- **flood damage reduction**   A large number of satellites can be used to estimate hydrological variables (snow cover, water elevation, rain rate, soil moisture, solar radiation, surface albedo, land cover, flood monitoring and surface temperature). The real strength of RS in flood mitigation is the supply of parameters that are difficult to measure for flood modelling, particularly data on land use and vegetation. This information is also required in the development of flood-risk maps.

Use of geographical information systems (GIS) in disaster reduction

GIS provides a key resource for disaster mitigation, combining physical data (slope, geology, soils and drainage) with human data (census, housing types and location of vulnerable people). GIS creates 'layers' of electronic data.

Applications of GIS in disaster reduction and management include:

- **hazard mapping**   Identifying the risk from natural disasters is an important requirement for mitigation and preparedness. Combining hazard maps with population and infrastructure allows hazard plans to be prepared. Figure 42 is a lava-flow hazard map for Hawaii. Each zone represents the likelihood of coverage by lava from the volcanoes. Zones 1–3 are limited to the active volcanoes of Kilauea and Mauna Loa. Zone 1, the most hazardous, includes the summits and rift zones of Kilauea and Mauna Loa where the vents have been repeatedly active and where lava flows will originate in the future.

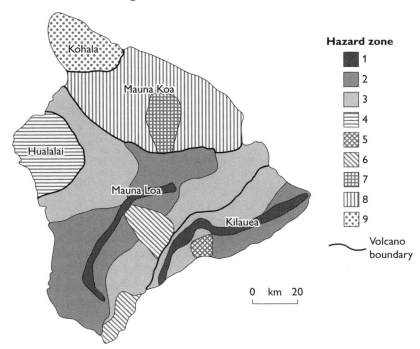

Figure 42 Hazard map of lava flow for Hawaii

For response, it is necessary to identify passable roads and the locations of emergency services, refugee camps and feasible transportation routes.

- **support for simulation and modelling**   Mathematical simulation and modelling disaster processes is the main procedure in forecasting and warning, as well as in impact assessment. GIS is widely used to prepare data for mathematical models. The ability of GIS to process complex spatial information has helped to produce more complex models that represent the disaster scenarios in more detail.
- **GIS in planning**   GIS has become a prominent tool for city and infrastructure planning. There are examples of its use in land management response and recon- struction. More specifically, GIS is used by industry in securing pipelines such as gas, water and electricity. It has also been used in selecting locations for resettle- ment after a disaster.

Using technology in communication

Communication is an essential component in all aspects of disaster management, from mitigation, preparedness and response to recovery at global, national and local levels. The enormous possibilities offered by the internet and the advent of wireless telecommunications have resulted in increased interest and opportunities. It is expected that global communication will further decrease in cost with the deploy- ment of big and small low earth orbit (LEO) and medium earth orbit (MEO) satel- lites in the near future.

Mini case study: Kyoshin Net
After the Kobe earthquake, the National Research Institute for Earth Science and Disaster Prevention (NIED) deployed 1000 strong-motion accelerometers throughout Japan. This network is called the Kyoshin Net. The average distance between stations is 25 km. During an earthquake, P and S wave velocities are measured and logged at each site. Data are then sent to the local municipality via a modem. The munici- pality can use the information for local emergency management and response.

The same information is sent to the control centre at NIED. These records are collated and then published on the internet. The centre also maintains a strong- motion database and site information for scientific studies and engineering applica- tions.

Mini case study: how to predict an eruption
Changes in the direction of seismic waves around active volcanoes could help warn of an impending eruption almost a year before the event. Researchers in New Zealand claim to have identified a telltale signature for Mt Ruapehu, which blew its top in 1995. The findings could be applied to other volcanoes. Seismometers installed around the mountain detect changes in wave direction as magma pressure builds up in the volcano. This means that the 'stress' of the volcano can be measured *in situ*.

Is technology the answer?

- It is important to provide a framework in order for technologies to be useful at national and local levels. For example, hazard maps or risk maps should be disseminated and used by the communities for which they were developed. This has to be achieved through local governments as part of their legislation.
- Are technologies available to communities? There will be instances when it is not practical or feasible to have the technical capability locally (e.g. in NICs and LEDCs).
- Standardisation of information is necessary so that data generated by one organisation can become input information for other organisations.
- Information should be made accessible to the general population and the research community. The USA has taken the lead in this respect, while Japan has been slow to implement data accessibility.

The future

The view that disasters are temporary disruptions to be managed solely by humanitarian response or that their impacts will be reduced only by technical interventions has been replaced by the recognition that they are intimately linked with sustainable development activities in the social, economic and environmental fields. So-called 'natural' disasters are increasingly regarded as one of a number of risks that people face. Others include epidemics, economic downturns, lack of food/clean water, unemployment and insecurity. Where some of these risks are compounded, impacts of disasters are often exacerbated.

The UN International Strategy for Disaster Reduction has identified some key priorities for the future of disaster and risk management:

- There is a need for disaster and risk reduction to be an essential part of the broader concerns of sustainable development. For example, risk assessments and vulnerability reduction measures should be taken into account in different fields, such as environmental management, poverty reduction and financial management.
- Not all current development practices reduce communities' vulnerability to disasters. Indeed, ill-advised and misdirected development practices may actually increase the risks.
- Political commitment by public and private policy makers and local community leaders, based on an understanding of risks and disaster-reduction concepts, is fundamental to achieving change.
- National and local authorities bear the main responsibility for the safety of their people. However, it is the international community's duty to advocate policies and actions in developing countries that pursue informed and well-designed disaster risk reduction strategies. It must also ensure that its overseas programmes reduce, and do not increase, disaster risks.

Generalisation 4: Global trends in hazards and hazard management

There are three parts to this generalisation:

- Global trends in hazard **occurrence**, and the **impacts** that hazards make on property and infrastructure (**economic costs**) and deaths and human welfare (**social costs**).
- Global trends in hazard **management**, in particular, assessing how science and technology can improve the **prediction** of hazards and the extent to which this may reduce impacts.
- Future management strategies, in particular, assessing how the **causes** of hazards might be managed (**modifying the event**) as opposed to the current strategies of **managing their effects** (**modifying vulnerability or loss**).

These three ideas are examined globally and nationally. Countries at various stages of development are compared.

Global trends in the occurrence and impact of hazards

Figure 43 summarises four key trends in natural disasters. It is based on statistics gathered by CRED over the last 30 years.

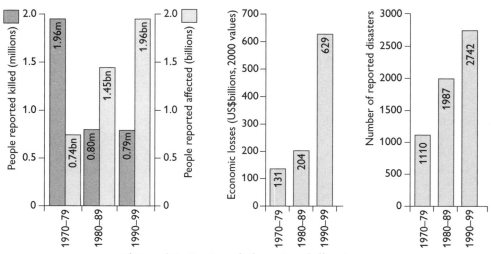

Figure 43 Key trends in natural disasters

- The number of people reported killed initially dropped dramatically and then levelled off — the statistics for 1980–89 and 1990–99 are similar.
- The number of people reported affected by hazards/disasters (e.g. injured or with loss of livelihood) has increased each decade.
- The economic losses have grown exponentially.
- The number of reported disasters has grown significantly by about 800 each decade.

Disaster statistics — a health warning

Is the source of the data reliable?

Governments report disaster statistics to UN agencies. These quasi-official statistics are supplemented and crosschecked by designated groups that also monitor reports from the media (internet data are important) and from NGOs (e.g. Oxfam) working at the 'front line'. The pre-eminent authority is CRED, supported by the World Bank and insurance companies such as Munich Re, Swiss Re and Lloyds of London.

How good are disaster statistics?

Problems associated with disaster statistics include the following:

- There is neither a universally agreed definition of a disaster nor a universally agreed numerical threshold for disaster designation. Reporting disaster deaths is controversial, since it depends on whether direct (primary) deaths or indirect (secondary) deaths from subsequent hazards or associated diseases are counted. Location is significant because local or regional events in remote places are under recorded.
- Declaration of disaster deaths and casualties may be subject to political bias. The Boxing Day tsunami was almost completely ignored in Myanmar (Burma) but perhaps initially overstated in parts of Thailand, where foreign tourists were killed and reports were played down to protect the Thai tourist industry.
- Statistics on major disasters are difficult to collect, particularly in remote rural areas of LEDCs (e.g. Kashmiri earthquake, 2005) or in densely populated squatter settlements (e.g. Caracas landslides, 2003–04).
- Time-trend analysis (interpreting historical data to produce trends; see Figure 43) is difficult. Much depends on the intervals selected and whether the means of data collection have remained constant. Trends can be upset by a cluster of mega-disasters, as happened in 2005–06.

The number of deaths from disasters

Globally, the annual number of deaths from disasters seems to have fallen, or at least levelled off. Ninety per cent of such deaths occur in developing countries. Twenty-four of the 49 least developed countries (LDCs) face high levels of disaster risk, with six countries experiencing between two and eight major disasters during each of the past 15 years.

The decline in disaster-related deaths should be set against the backdrop of rising hazard numbers and increasing numbers of vulnerable people. Figure 44 (the **risk disc**) explains the reasons for the decline in deaths in terms of disaster preparedness, disaster mitigation (hazard proofing), disaster response and disaster recovery.

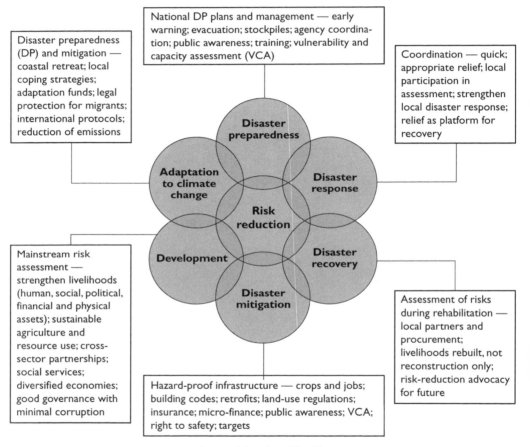

Figure 44 The risk disc

In 1994, in Yokohama, at the mid-decade conference for the International Decade for Natural Disaster Reduction, an important watershed was reached. Delegates argued for a less **top-down** technocratic fix and a more **bottom-up** approach, using NGOs to mobilise local communities (particularly in large cities) in strategies to improve their resilience to disaster. It is the combination of these two approaches that has led to the reduction in deaths.

Mini case study: responses to the savage storms of 2004

In 2004, a series of hurricanes crisscrossed the Caribbean. The responses of countries were largely dependent on the level of technology and the degree of political stability.

In Cuba, the response successfully combined the political efficiency of a totalitarian state with sound technology and outstanding community preparedness. Civil defence is now part of national security and during Hurricanes Charley and Ivan only four people were killed. Cuba has a world-class meteorological institute that produces excellent computer models and reaches people through all types of media.

There is education on the dangers of hurricanes and the systems are well consolidated at national, regional and local government levels.

In Jamaica, preparedness paid off. Since Hurricane Gilbert, there have been community disaster response officers and plans in each parish. Factors include:
- good maps showing resources and the homes of the most vulnerable people
- trained and well-equipped Red Cross emergency helpers, wardens, drugs, refuge shelters (e.g. in all schools)
- voluntary, annual Gilbert anniversary practice drills

Fourteen people were killed in Hurricane Ivan.

In the Dominican Republic, the focus was on wind hazard, yet the main problem in 2004 was flooding (100 mm in 24 hours). Positive factors now in place include:
- delivery of maps, and visits from Red Cross outreach workers
- a new civil defence system, based in churches

A problem was a lack of short-wave radios and other key equipment. Twenty people drowned during Hurricane Jeanne.

Haiti is the poorest Caribbean country. A political vacuum led to lack of local government organisation. Extreme poverty has caused a lack of resources — for example, at the national meteorological centre there is lack of local systems. The people are poorly educated, so it is difficult to develop communications effectively.

During Hurricane Jeanne, up to 3000 people died from mudslides caused by torrential rain,

Number of people affected by hazards and disasters

The number of people affected annually by hazards and disasters has increased considerably (Figure 43). 'Being affected' includes loss of home, crops, animals or livelihoods, or decline in health for a designated period of time (often 1–3 months). On average, around 200 million people per year are affected by disasters. The vulnerability progression (Figure 45) is responsible for the concentration of affected people in LEDCs.

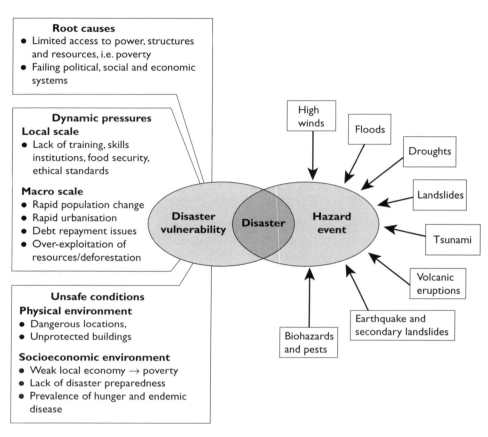

Figure 45 The vulnerability progression

There are a number of interlinking factors that have led to the increasing number of people, particularly in LDCs, being vulnerable to disaster (Table 10).

Table 10 Factors leading to increasing vulnerability

Factor	Impact on vulnerability	Example
Population growth and change	The world's population should stabilise at 9 billion in 2050; the growing proportion of the very young and very old will increase the number of people affected by disasters	Growing concern over the vulnerable aged in Japan and Florida, USA
Land pressure	Land pressure leads to the occupancy of areas of high risk, the cutting down of forests to provide farmland and the destruction of mangroves to allow coastal development	In Bangladesh, 85% of the population depends on subsistence farming; millions live on disaster-prone floodplains

Factor	Impact on vulnerability	Example
Urbanisation	The rapid rate of urbanisation in the developing world is a key factor in the growth of vulnerability, particularly of low-income families living in squatter settlements/slums; squatter settlements occur in zones of high risk; slums grow up in inner-city tenements; many megacities are in multiple-hazard zones	Caracas landslides, 1999; Mexico City earthquake, 1985 which demolished the slum areas; the garbage dump mudslides of Manila City
Political change	This can lead to destabilisation; corruption or war can exert extra land pressure and lead to desperate responses by rural migrants	Darfur, Sudan; Haiti between governments during Hurricane Jeanne
Economic growth	This leads to development of sophisticated structures such as bridges; it can also lead to loss of forests, soil degradation and overuse of resources, all of which increase vulnerability	Texas and Atlantic coasts in USA are vulnerable to hurricane damage, e.g. bridges that link barrier islands
Globalisation	This has led to TNCs being major world players who exploit natural resources	Timber companies in the Philippines; tropical storm Debbie led to major flash flooding and siltation in St Lucia, where watersheds had been deforested by banana plantations
Technological innovation	Flood barriers have been built along rivers and coasts, but it is usually too costly to prepare for even a 1-in-200-year event; false security may encourage building in unsafe areas	In New Orleans, the river flood-control banks had not been strengthened to cope with a disaster of Hurricane Katrina's proportions
Development gap and the occurrence of poverty	As the development gap widens between LDCs and the rest of the world, the problem of debt servicing means that many countries have no money available for disaster management	DFID is now providing grants (e.g. to Bangladesh) specifically for flood action and cyclone action plans
Climate change	Human-induced global warming is causing long-term climate change that could lead to deforestation or desertification; short-term extreme weather events such as droughts/fires and floods occur; increased ocean warming may be spawning more hurricanes	The 1989 and 1997 El Niño events were felt around the world; European floods, 1999–2000, 2002, and European droughts in 2003 and 2006 could be related to climate change

Megacities in hazard areas and their projected growth are shown in Table 11. Some vulnerable cities in LEDCs are expected to grow by more than 5 million.

Table 11 Megacities in hazard areas

City/conurbation	Population, 1996 (millions)	Projected population, 2015 (millions)	Hazard(s)
Tokyo–Yokohama	27.2	28.9	Earthquakes, cyclones
Mexico City	16.9	19.2	Earthquakes, floods, landslides
São Paulo	16.8	20.3	Landslides, floods
New York*	16.4	17.6	Winter storms, cyclones
Mumbai	15.7	26.2	Earthquakes, floods
Shanghai	13.7	18.0	Floods, typhoons
Los Angeles*	12.6	14.2	Earthquakes, landslides, wildfires, floods, smogs
Calcutta (Kolkata)	12.1	17.3	Cyclones, floods
Buenos Aires*	11.9	13.9	Floods
Beijing	11.4	15.6	Earthquakes
Lagos	10.9	24.6	Floods
Osaka*	10.9	10.6	Earthquakes, cyclones, floods
Rio de Janeiro*	10.3	11.9	Landslides, floods
Delhi	10.3	16.9	Floods, heat and cold waves
Karachi	10.1	19.4	Earthquakes, floods
Cairo–Giza	9.9	14.4	Floods, earthquakes
Manila	9.6	14.7	Floods, cyclones, earthquakes
Dhaka	9.0	19.5	Floods, cyclones, storm surges
Jakarta	8.8	13.9	Earthquakes, volcanoes
Tehran	6.9	10.3	Earthquakes
* = NIC or MEDC			

Economic losses from disasters

Economic losses from disasters have grown exponentially (Figure 43). Comparing 1980–89 with 1990–99 shows that economic losses tripled. Losses are growing at a far greater rate than the number of disasters. Munich Re, the reinsurance giant, looked at the trend of economic losses and insurance costs over a 50-year period (Figure 46).

Rising trends can be identified. However, mega-disasters such as Hurricanes Andrew and Katrina lead to huge rises for a single year, resulting in a **fluctuating** trend. The trend of insured losses is rising less rapidly than that for total economic losses. This is a reflection of:

- the spatial concentration of hazards
- their realisation into disasters in LEDCs

- the poverty-stricken underclasses in MEDCs not being able to afford the high cost of insurance in **hazard-prone zones** (e.g. some districts of New Orleans)

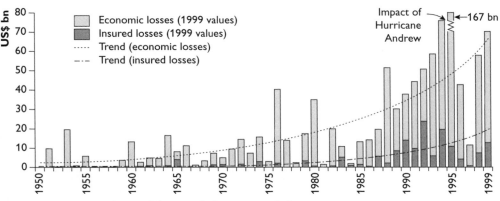

Figure 46 Trends in economic losses, 1950–2000

There is a tendency to over-emphasise the economic damage and losses experienced in MEDCs. In **absolute terms**, there is no doubt that because of the value of their economies, the sophistication of their installations and infrastructure, and the scale of domestic insurance claims, the amounts involved will be large. Figure 47(a) compares insurance losses of recent **mega-events**. In human costs (Figure 47(b)), the rank order is different, reflecting the MEDC/LEDC divide. This is often expressed by the simplistic statement 'MEDCs experience the greatest damage (economic cost), whereas LEDCs experience the greatest number of deaths (social cost)'.

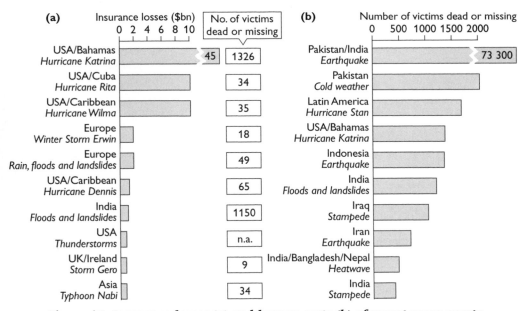

Figure 47 Insurance losses (a) and human costs (b) of recent mega-events

In reality, the situation is economically more complex. In *relative* terms, economic damage from natural disasters tends to be higher in LEDCs, mainly because of their high dependency on one or two cash crops or on tourism. The damage forms a high percentage of their annual GDP (Figure 48).

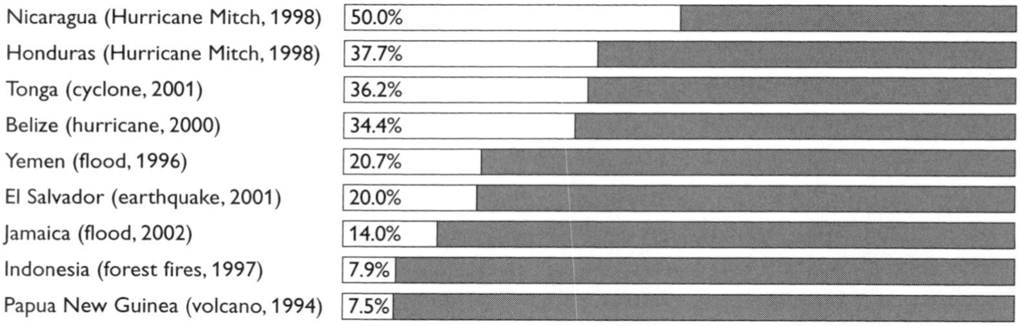

Nicaragua (Hurricane Mitch, 1998)	50.0%
Honduras (Hurricane Mitch, 1998)	37.7%
Tonga (cyclone, 2001)	36.2%
Belize (hurricane, 2000)	34.4%
Yemen (flood, 1996)	20.7%
El Salvador (earthquake, 2001)	20.0%
Jamaica (flood, 2002)	14.0%
Indonesia (forest fires, 1997)	7.9%
Papua New Guinea (volcano, 1994)	7.5%

Figure 48 Percentage of GDP represented by losses from some natural disasters, 1994–2002

MEDCs have more diverse economies that are better able to withstand natural disasters. For example, Munich Re estimated that between 1994 and 2003, Japan suffered $166 billion of economic damage from natural disasters. This was only 2.6% of its GDP.

In conclusion, the rate of economic losses is increasing much faster than the occurrence of disasters, largely because of the growing economies of RICs and NICs, especially in Asia.

Number of reported disasters

The number of reported disasters has grown by about 800 each decade since 1970 (Figure 43). When profiling disasters, key features are the frequency of occurrence, and the magnitude of the event. The trends in frequency and magnitude affect their impact in terms of economic and social costs.

Figure 49 looks at the rising trends by major hazard type. Earthquakes, storms (including hurricanes) and floods comprise around two-thirds of all major disasters. In recent years, drought has become more widespread and significant, affecting millions of people (10%). Other hazards such as volcanoes (2%), tsunamis (1%) and avalanches (1%) are 'rare', but can be devastating. Figure 49 shows that the rising trend is a result of hydrometeorological hazards (floods, storms and droughts); earthquakes show fluctuations (long **timescale variations**) but no rising trend.

Consider the incidence of tectonic hazards. Timescale variations can be explained by spatial clustering along highly mobile plate boundaries. Earthquakes, and their associated volcanic activity, frequently occur in series, such as those currently occurring along the Sunda fault off the coast of Sumatra, where the Indian plate is subducted beneath the Burma plate. Scientists liken the Sunda fault to a zip that is gradually

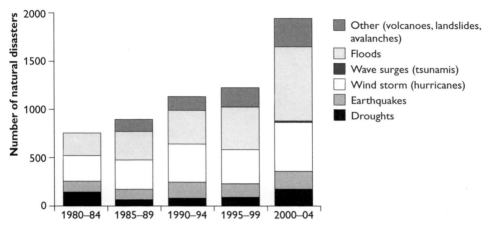

Figure 49 Frequency of natural disasters, 1980–2004

coming undone as each earthquake transfers stress along its length. Recently there have been two major earthquakes in Sumatra (2005) and one in Java (2006), with two generating tsunamis as a sudden change occurred in the sea floor.

Arguments put forward to explain the increase in the number of reported meteoro-logical events are usually associated with climate change. Tables 12 and 13 show the ways in which the Intergovernmental Panel on Climate Change predicts how a more extreme climate (resulting from global warming) will have a hazardous impact on people and environments. Global warming is classified as a *context* hazard.

Table 12 Simple extremes resulting from projected climate change

Projected changes during twenty-first century (90–99% chance)	Possible impacts
Higher maximum temperature; heatwave over nearly all land areas (very likely)	• Increased deaths of elderly and urban poor (Paris, 2003) • Increased heat stress of livestock/wildlife • Shift in tourist destinations • Huge demand for air-conditioning (failure of electricity supply in California, 2006)
Higher minimum temperature; fewer cold days, frost days and 'cold waves' over nearly all land areas	• Decreased deaths from hypothermia • Changing patterns of crop growth • Extended range and activity of some pest and disease vectors • Reduced demand for heating energy • Changing winter tourism patterns
More intense precipitation events (storms) over many areas	• Increased flood, landslide, avalanche, mudslide and debris-flow damage • Increased soil erosion • Increased flood runoff, so more pressure on government, private flood insurance and disaster relief schemes

Table 13 Complex extremes resulting from projected climate change

Projected changes during twenty-first century (66–90% chance)	Possible impacts
Increased summer drying and associated risk of drought in most mid-latitude continental interiors	• Increased risk to quantity and quality of water resources • Increased risk of forest fires • Increased subsidence/shrinkage of building foundations, especially in clay areas • Decreased crop yields
Increased tropical-cyclone-peak-wind intensity, mean- and peak-precipitation intensity	• Increased risk to human life, and of infectious diseases and epidemics • Increased coastal erosion and damage to coastal buildings and infrastructure • Increased damage to coastal ecosystems (e.g. corals and mangroves)
Intensified drought and floods associated with the El Niño–La Niña cycle, with teleconnections across many regions	• Decreased agricultural and range-land productivity (e.g. mid-west USA, flood and drought prone) • Decreased HEP potential, in some drought-prone regions (e.g. New Zealand)
Increased variability of Asian monsoon precipitation	• Increased flood and drought magnitude and damages in both temperate and tropical Asia (India, 2006)
Increased intensity of mid-latitude storms	• Increased risks to human life and health • Increased property and infrastructure losses

Projected climate change in coming decades is likely to alter the frequency and magnitude (intensity) of natural hazards:

- Changes in precipitation patterns, soil moisture and vegetation cover could lead to more frequent droughts, floods and subsequent landslides.
- Increased temperatures will warm the oceans, which could spawn more intense storms (category 4 or 5). In spite of much discussion about the frequency of hurricanes being related to global warming, most scientists link this more to oscillations, such as the El Niño southern oscillation. They note that hurricanes are minimal in the Atlantic in El Niño years, but at their worst in La Niña years. Current research suggests that seasonal oscillations between the Atlantic and Pacific influence the occurrence of storms.
- Global warming is leading to thermal expansion of the oceans. Rising sea levels are creating higher storm surges during hurricanes (Hurricane Katrina).
- Global warming is leading to more unpredictable climates with more extreme weather events such as the European floods in 2002 and European droughts in the summers of 2003 and 2006.
- Some scientists suggest that global warming is leading to more frequent and intense El Niño events, with 1997–98 being the strongest ever recorded. The El Niño–La Niña cycle led to numerous hydroclimatic hazard events (e.g. floods in Peru and drought/fires in Australia and Indonesia).

Another explanation for increased frequency of hydroclimatic disasters lies with increased environmental degradation caused by population pressure, which leads to deforestation and other land-cover loss. This can lead to flash flooding. Scientists are divided as to whether watershed deforestation and removal of land cover is the root cause of increased flash flooding in, for example, China, Vietnam and Bangladesh. Environmental degradation influences the effects of natural hazards by exacerbating their impacts.

Review

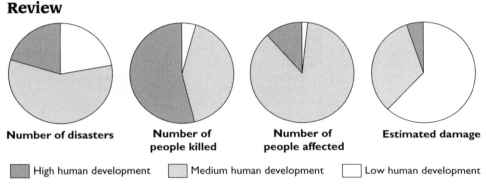

Number of disasters   **Number of people killed**   **Number of people affected**   **Estimated damage**

High human development   Medium human development   Low human development

Figure 50 Disasters related to human development

Figure 50 summarises how the trends discussed relate to levels of development. The incidence of reported natural disasters is much higher in poorer, less developed countries because they are located in tropical areas, in areas of monsoonal rainfall or in areas with highly mobile plate boundaries. Hazards cause disproportionately greater havoc in poor countries, which may lack the management resources and technology to cope. This has implications for the distribution of international disaster relief and assistance.

In some LEDCs, the development gap is becoming a '**disaster gap**' — physical and human factors combine in such a way that hazards become disasters.

Hazard prediction

The use of technology to improve **hazard prediction** is seen by some scientists as a key means to reduce vulnerability, i.e. impacts on people and their property.

The purpose of prediction is to allow people to respond at individual, community, local, regional, national and international levels. Accurate prediction can buy time:
- to warn people to evacuate
- to prepare for a hazard event
- to manage impacts more effectively
- to help insurance companies assess risks

- to prioritise government spending
- to help decision makers carry out cost/benefit calculations (e.g. for the construction of defences)

Prediction can increase understanding, because scientists can model and test their predictions and then compare them with reality.

Figure 51 summarises the importance of prediction.

When?	Where?	What?
When? • **Recurrence intervals** — an indication of longer-term risk • **Seasonality** — climatic and geomorphic hazards may have seasonal patterns, e.g. Atlantic hurricanes occur from June to November • **Timing** — the hardest to predict, both in the long term (e.g. winter gales) and the short term (e.g. time of hurricane)	**Where?** • **Regional scale** — easy to predict, e.g. plate boundaries, 'tornado alley', drought zones • **Local scale** — more difficult, except for fixed-point hazards, e.g. floods, volcanoes, coastal erosion • Moving hazards — extremely difficult, e.g. hurricane tracking	**What?** • **Type of hazard** — many areas can be affected by more than one hazard; purpose of forecast is to predict what type of hazard might occur • **Magnitude of hazard** — important in anticipating impacts and managing a response • **Primary vs secondary impacts** — some hazards have 'multiple' natures; earthquakes may cause liquefaction, volcanoes may cause lahars.

Hazard prediction

Why?	Who?	How?
Why? • **Reduce deaths** — by enabling evaluation • **Reduce damage** — by enabling preparation • **Enhance management** — by enabling cost–benefit calculations and risk assessment • **Improve understanding** — by testing models against reality • **Allow preparedness plans to be put in operation** — by individuals, local government, national agencies	**Who?** • **Tell all?** — fair, but risks over-warning, scepticism and panic • **Tell some?** — for example, emergency services, but may cause rumours and mistrust • **Tell none?** — useful to test predictions, but difficult to justify	**How?** • **Past records** — enable recurrence intervals to be estimated • **Monitoring (physical)** — monitored and recorded using ground-based methods or, for climatic and volcanic hazards, remote sensing • **Monitoring (human)** — factors influencing human vulnerability (e.g. incomes, exchange rates, unemployment); human impacts (e.g. deforestation)

Figure 51 Prediction

There are sometimes problems with the reliability of prediction technology. False warnings cost huge amounts of money (e.g. £30 million in Hawaii for each false tsunami warning) and lower confidence and credibility. People obeyed the first warning in the 2004 Florida hurricane season but were cavalier about subsequent ones, because the hurricane tracks were capricious.

Prediction relies increasingly on sophisticated technology, so it is expensive to resource. As a result of the technology gap, there is spatial inequality between LDCs/LEDCs and NICs/MEDCs.

Prediction technology: tectonic hazards

Earthquake

- Past magnitude and frequency data show risk areas and allow prediction of the probability of earthquakes. However, with such quick-onset events, timing cannot be precise.
- Seismic gap theory is useful for providing long-term prediction of areas likely to move in the future, i.e. mathematical modelling of the probability of an earthquake occurring.
- Short-term prediction a few hours before an event uses changes in groundwater levels or radon gas release at monitoring sites.
- Tilt meters and gravity meters, laser reflectors, strain meters, magnetometers and seismographs (to monitor foreshocks) are used to monitor ground dilation and rock-cracking just before the earthquake. Seismic computer systems process the data.
- Computerised GIS can produce a hazard-zone map that uses past information to record, for example, landslide potential. Such maps can be used in land-use planning.

Problem: the short onset time

Volcanic eruptions

- Physical processes can be monitored for changes that signal an impending eruption.
- Observation boreholes are used to monitor changes in temperature, composition of hot-spring waters, and sulphur dioxide in underground gases.
- Tilt meters measure mountain (land) swelling as magma rises to the surface.
- Aircraft and geostationary satellite remote sensing measure thermal radiation, gas and ground movements.
- MEDCs (e.g. Japan and USA) have volcano observatories that act as nerve centres for monitoring stations. Each has tilt meters, magnetometers, seismometers etc. Twenty per cent of the Earth's volcanoes are monitored in this way; GPS should mean that more are monitored in the future.

Problem: prediction is useful but may be made too far ahead, resulting in false alarms.

Tsunamis

- Ocean-wide **warning systems** alert areas at risk within 1 hour of a possible occurrence.
- Tsunami hazards-reduction utilising systems are based on rapid warning via satellite communication links and may provide cheap warning systems.
- Receptors on the seabed and buoys at the sea surface are used to detect tsunamis and give a warning just before onset.

Problem: not all earthquakes (even shallow, high-magnitude ones) lead to tsunamis.

Prediction technology: hydroclimatic or meteorological hazards

Tropical cyclones

- Coastal areas at risk are protected by warning systems that monitor development. They forecast intensity and tracks to permit sheltering and evaluation.
- The National Hurricane Center in Miami, Florida uses data from geostationary satellites, buoys and land/sea recording centres to monitor systems. Data are then beamed via satellites for analysis by super-computers.
- Reconnaissance aircraft are used to provide detailed hurricane profiles.
- Analogue computer models of atmospheric processes and statistical models of hurricane activity are used to make predictions.
- Hurricane-risk maps inform land-use mapping.

Problems: endless tracking, but the systems are capricious; it takes time to evacuate people.

Floods

- GIS is used to collect meteorological, discharge and tidal data.
- ICT is used to disseminate flood warnings more efficiently to vulnerable people.
- Detailed flood-risk mapping (using GIS) is used to inform insurers and land-use planners.

Other hazards

Drought   Computer data analysis is used to predict long-term onset of droughts. Satellite imagery is used to measure eco-stress to assess onset of drought.

Tornadoes   Satellite technology, combined with Doppler radar, is used to predict likely conditions for tornado formation.

Wildfire   Infrared photography in satellites is used to analyse the distribution of fires to improve monitoring. Night-time thermal emission anomalies give the temperature and size of fires.

Snow avalanches   Computer forecasting of avalanches from meteorological (temperature and precipitation) data and snow-pit stability tests is linked to central analysis centres. Avalanche tracks are analysed by computer.

Risk reduction

High-tech equipment is not the sole answer to predicting hazard events. The prediction has to be converted into action to reduce the vulnerability of the population. To increase individual, community and national **resilience** to hazards, the physical analysis of the likely impact of hazard events must not be separated from social and economic development.

Review

The Boxing Day 2004 tsunami showed that lives could have been saved by the installation of an early-warning system, which would have predicted the tsunami for everywhere except Sumatra (too near the epicentre). However, the technology must be combined with programmes to reduce the risk for local communities.

Modifying the hazard event

Causes of hazards are either primary (e.g. slope failure that triggers a landslide) or root causes induced by humans. Root causes are being managed increasingly as part of integrated risk-reduction strategies to modify vulnerability; the management of primary causes is for the future. Managing the physical processes involves modifying the environment in some way, so that physical exposure to the hazard is prevented. Factors that influence the degree of modification possible include hazard size and scale, speed of onset and the reliability and cost of existing technology.

There are ethical issues in the more experimental approaches (humans controlling nature) and there is also the risk of adverse ecological and environmental consequences.

Tectonic hazards

- Modifying and managing the causes of **earthquakes** is not feasible, although experiments with lubricating the San Andreas fault plane have been conducted to try to prevent the jarring that results from sudden slippage.
- **Volcanic eruptions** cannot be prevented. Attempts have been made to control lava flows by bombing fluid lava or by using artificial barriers to divert it (e.g. in Hawaii). On Heimaey in 1973, seawater from the harbour was used to chill the advancing lava from the Eldfell eruption.
- The causes of **tsunamis** are impossible to control.

Geomorphic hazards

- For **avalanches**, snow packs can be stabilised. Artificial release occurs at predetermined times when ski runs and highways are closed. The snow pack is released safely in several small avalanches, rather than allowing a major threat to build up. In parallel with monitoring snow stability, small explosives are used to trigger controlled avalanches, particularly in the starting zone. Where highways are endangered, the military may trigger avalanches with field guns. There are defence structures to control the development of wet-snow avalanches; powder avalanches are difficult to manage.
- **Landslides** and other slumping can be prevented by slope-defence methods such as excavating and filling, draining, re-vegetation, restraining structures and chemical stabilisation (e.g. in high-risk urban areas of Hong Kong).

Meteoroclimatological hazards

- Suppression of **severe storms** remains a dream. A 10% reduction in wind speed could downgrade a tropical cyclone to tropical storm status and, therefore, perhaps reduce damage by 30%. In the 1960s, Project Storm Fury aimed to introduce freezing nuclei into the ring of clouds around storm centres to stimulate the release of latent heat to lower the maximum temperature in the hot core. Cloud seeding has been used for suppression of hail in severe storms. Current research is concentrated on attempting to cool down the oceans that generate hurricanes by covering the surface with biodegradable oil, and on how to cause hurricanes to change course and veer away from densely populated areas.

- In theory, **droughts** can be prevented by artificial stimulation of rainfall by silver iodide seeding. However, the clouds must have natural precipitation potential.
- Causal management of some **floods** is feasible, largely by attenuating the lag time — using dams and reservoirs, and afforestation of the upper catchment.

Review

It is beneficial to manage the effects of hazards by:

- **addressing root causes** — ensuring a strong, ethical government; attempting to 'make poverty history'
- **reducing pressures** — community preparedness; education and training to develop a safety culture; developing community resilience
- **reducing hazards** — hazard mitigation, e.g. use of technology to provide flood controls
- **achieving safe conditions** — hazard protection and prevention; land-use planning; hazard-resistant structures; warning systems; contingency planning

Questions
&
Answers

This section begins with general guidance on how to write quality hazard essays. This is followed by two responses to the same essay question. One answer is a borderline grade-A response; the other is borderline grade E. The threshold for an A-grade answer is 44–46 marks out of 60. The threshold for an E-grade answer is usually 27–30 marks.

Examiner's comments

Candidate responses are interspersed with examiner's comments, preceded by the icon 🄴. The comments indicate how each answer would have been graded in the actual exam and give suggestions for improvement.

The marks for the two example essays are summarised below.

	A-grade essay	E-grade essay
D	6	7
R	12	6
U	12	6
C	6	4
Q	8	5
Total	44	28

How to write quality hazard essays

Deconstructing the title

You have to be able to understand the **command word**. For an A2 essay, the following command words are important:

- **Discuss** — investigate, giving evidence for and against the statement
- **Examine** — look closely into the statement giving detailed support
- **Evaluate/assess** — make a judgement, supported by evidence
- **Explain** — clarify, interpret and account for; give reasons for

Extract the **key words**. These could be:

- **topic** key words (e.g. tectonic hazards)
- **place** and **scale** key words (e.g. LEDC/MEDC or national/local)
- **issue** key words (e.g. factors or role)

A deconstructed title is shown in Figure 52.

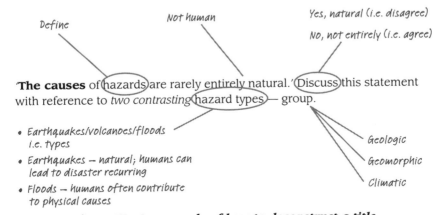

Figure 52 An example of how to deconstruct a title

Planning

After deconstructing the title, you should plan the essay. In the exam, you have about 10 minutes to do this. Most people think round a title using a spider diagram or similar. However, the ideas then have to be classified and sequenced.

Figure 53 is a plan for the deconstructed title given in Figure 52.

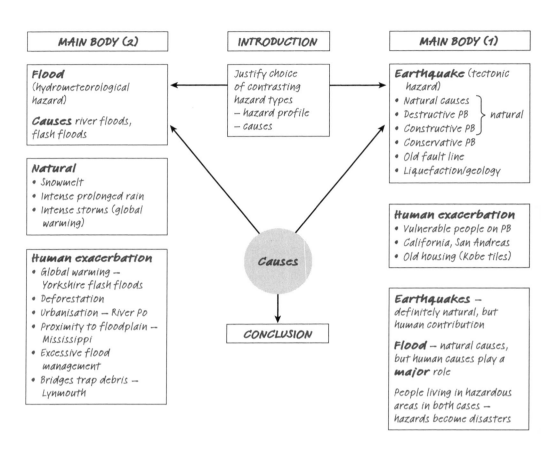

Figure 53 Planning your essay

Figure 54 shows an orderly plan in response to the title: 'The main aim of hazard management should be to reduce the effects of hazards, not manage their causes'. Discuss.

The student has devised a neat visual scale across a range of hazards to explore the concept of a spectrum of management from 'manage causes' to 'reducing effects'. This is followed by clear ideas about what to put into the introduction, the main body of the essay and the conclusion.

It is essential that you practise planning answers to a range of titles because a good plan is usually the key to a good essay. In the exam, write the plan inside your answer book and do not cross it out. If you run out of time, the plan gives the examiner a clear idea what your design was and you may gain some credit for this.

Tip

Avoid planning case study by case study. This leads to descriptive essays full of death and destruction.

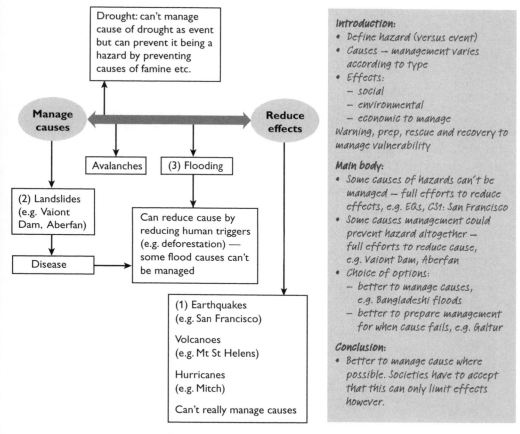

The figure contains the following text boxes:

Drought: can't manage cause of drought as event but can prevent it being a hazard by preventing causes of famine etc.

Manage causes

Reduce effects

Avalanches

(3) Flooding

(2) Landslides (e.g. Vaiont Dam, Aberfan)

Can reduce cause by reducing human triggers (e.g. deforestation) — some flood causes can't be managed

Disease

(1) Earthquakes (e.g. San Francisco)

Volcanoes (e.g. Mt St Helens)

Hurricanes (e.g. Mitch)

Can't really manage causes

Introduction:
- *Define hazard (versus event)*
- *Causes — management varies according to type*
- *Effects:*
  - *social*
  - *environmental*
  - *economic to manage*
Warning, prep, rescue and recovery to manage vulnerability

Main body:
- *Some causes of hazards can't be managed — full efforts to reduce effects, e.g. EQs, CS1: San Francisco*
- *Some causes management could prevent hazard altogether — full efforts to reduce cause, e.g. Vaiont Dam, Aberfan*
- *Choice of options:*
  - *better to manage causes, e.g. Bangladeshi floods*
  - *better to prepare management for when cause fails, e.g. Galtur*

Conclusion:
- *Better to manage cause where possible. Societies have to accept that this can only limit effects however.*

Figure 54 Planning a hazard essay

Exploring the mark scheme to maximise performance

For this exercise, you need to revisit the generic (DRUCQ) mark scheme (p. 6) and some of the question-specific mark schemes.

Writing the introduction

There are four elements required to write a successful introduction:

(1) *A clear statement of the question, problem, issue '*

Discuss/dissect the title before launching into definitions of the word 'hazard'. Write in the 'third person' (avoid 'I will...'). Check command words. Do not just repeat the title. Do not prejudge the outcome.

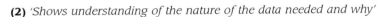

(2) *'Shows understanding of the nature of the data needed and why'*

You should make a brief mention of sources, the range of data, possible bias — but do not write a long methodology. Avoid setting up key concepts or key questions. The structure may not be by case study but by theme, for example scale or success.

(3) *'Refers to a potential range of scales and/or locations'*

Justify your choice of case studies — do not just list them. Range could mean size, impact, economic development or success. Do not waste time drawing a world map and locating examples on it. A model (e.g. development spectrum) *may* help. Do not introduce a model and then not use it.

(4) *'Accurate definitions of terms'*

Use definitions from established textbooks or reputable websites, not just from a dictionary or Wikipedia. Credit the source if possible.

Example introduction

According to John Whittow, 'A natural hazard is a perceived natural event which threatens both life and property'. However, Hewitt and Burton give a more precise definition — the event 'is a hazard when it causes US$50 000 of damage, over ten deaths or clear social disruption'. These definitions will need to be considered when examining different roles in the response to a hazard. **(1) (4)**

There are three different types of hazard (geomorphological, climatic and tectonic) and four main ways in which people respond to them. These are: do nothing, try to manage the causes and effects of the hazard, try to forecast the event in order to prepare and, lastly, engineering for protection and prevention purposes.

Four features of the specific hazard affect the method of response used: type, location, magnitude and frequency. The economic wealth of the country is also important. **(2)**

The main players in responding to hazards, i.e. the people involved, are the government, relief workers, insurance companies, individuals working as a community group and emergency services. **(4)**

Hurricane Katrina will be used as a topical climatic hazard in an MEDC, which prompted a huge international response, and contrasted with annual flooding in the LDC Bangladesh and the rare occurrence of a tornado in Birmingham in July 2005. The huge tsunami of December 2004 will be contrasted with that of Chimbote, Chile in February 1996. These provide a variety of scales and roles of people. The Parks model will be used to assess the groups' roles before, during and after the event. **(3)**

The above example is perhaps a little long. However, to earn the 10 marks the introduction needs to be solid — perhaps two-thirds of a side. When students launch straight into hazards without unpicking the title, they invariably drop a grade by losing the marks available for the introduction. They also often fail to focus on the question.

In answer to the question 'Explain why some hazards are more predictable than others', you should avoid the following type of introduction:

> In order to answer this question fully I will have to break down the question. I will start by defining a hazard, a natural event and a natural disaster. The question is asking me to explain, which means giving a thorough answer and backing it up with examples and proof. The word predictable means knowing something is going to happen. For my case studies, I am going to use Nevada del Ruiz, Kobe and the avalanches of Norway.

What is needed here is an explanation of the importance of prediction in preventing hazards becoming disasters. This should be followed by arguments that look at prediction in terms of 'when', 'where' and 'how large' for a range of hazard types in contrasting locations.

Research

Research involves getting together a range of relevant case studies (in depth) and examples (three or four sentences) to inform the question. Range means looking beyond just earthquakes or even just tectonic hazards. Getting the balance between depth of case study and breadth of argument is difficult. Well-drawn conceptual diagrams, maps and comparative factual tables will enhance your answer. This section must show evidence of your reading, with precise facts and figures about the case studies you have chosen.

Understanding

Showing understanding of, or applying knowledge to, the question is best done by concept, rather than by case study. It is also important that ongoing evaluation is included at the end of each paragraph. Try to make sure that each paragraph makes a particular argument, well supported by evidence.

Conclusion

The conclusion is frequently neglected. It should be meaningful and tie the essay together. You need to make it clear where the conclusion starts (e.g. 'In conclusion...'). The conclusion is worth 17% of the total marks, so you need to write about three-quarters of a side.

> ### Example conclusion
> In conclusion, it can be seen that some hazards are more predictable than others. In terms of 'where', fixed hazards such as volcanoes and river or coastal

flooding are predictable in terms of their location. At a **macro scale**, it could be argued that 99% of all tectonic hazards take place along plate boundaries and that hurricanes are confined to tropical areas. However, as Hurricane Katarina (Brazil) showed, new locations can occur as a result of changing conditions. Flash floods are unpredictable spatially (Thirsk, 2005).

'When' is much harder to predict. The arrival of some larger-onset hazards such as drought and large-river flooding can be predicted, but other hazards are difficult to pinpoint, in spite of extensive monitoring. In particular, hurricanes can slow down or 'loop-the-loop', so times of landfall are unpredictable. This was shown by Hurricane Charley (2004). Volcanoes and earthquakes can be declared imminent, but the precise timing may be several days or even months out.

Magnitude is also difficult to predict, particularly with a tsunami, which is a secondary tectonic occurrence. It depends not only on the magnitude of the earthquake on the Richter scale, but also on the configuration of the ocean with reference to the epicentre.

Putting together these factors, it can be seen that nearly all hazards are unpredictable. It is difficult to suggest one type, or one group, that is more predictable than others. It depends on which aspect of prediction is being considered.

Quality of written communication

Quality of written communication is important. Your essay structure should provide a linked sequence of analysis and discussion that maintains the argument. Command words such as 'discuss' or 'to what extent' require you to develop a mature, evaluative style.

It is important to develop good standards of spelling, particularly of geographical terminology and technical words. Spellings such as 'tec**h**tonics', 'volcano**e**' and 'ha**zz**ards' are to be avoided.

In general, essays are now marked online. This makes it particularly important that you use paragraphs of appropriate length to signpost your ideas.

So the advice is: make good use of the mark scheme. A badly written answer with limited introduction and conclusion can reduce your score by about 25 marks, which is the difference between unclassified and a good grade A.

Examples of essay titles

Generalisation 1: Causes of different types of hazard

- With reference to *one* group of hazards, explain why knowledge of physical processes is necessary to understand their occurrence and impacts.

- Although complex, the physical processes that cause meteoroclimatic hazards are well understood. Discuss.
- The severity of impact of a hazard results from its physical causes. Discuss.
- The causes of hazards are rarely entirely natural. Discuss this statement with reference to examples from *two* contrasting hazard types.
- Explain how physical geography has made some parts of the world more hazardous than others.
- Discuss the view that the causes of one group of natural hazards owes more to human factors than to physical factors.
- To what extent can hazards be classified by their causal processes?
- To what extent does an understanding of plate tectonics explain how the variety of tectonic hazards can occur?

Generalisation 2: Spatial variations in the impact of natural hazards

- Explain why some regions and countries are zones of multiple-hazard impact.
- Explain why *either* the environment *or* the economic effects of hazards vary spatially.
- How true is it to say that natural hazard events cause more damage in MEDCs and more deaths in LEDCs?
- The frequency and magnitude of natural hazards are the main factors that explain the spatial variation of their impact. Discuss.
- The impact of hazards is always greater in a multiple-hazard zone. Discuss.
- Analyse the factors that lead to spatial variations in the social and economic impacts of hazards.
- With reference to one group of hazards, compare their spatial impacts in LEDCs and MEDCs.
- Hazards are inevitable, but the damage caused is not. Discuss this view with reference to contrasting hazard types.
- Countries most at risk from disasters caused by natural hazards tend to be poor. Discuss.
- Evaluate the view that the distribution and density of population are the key factors in influencing the spatial variations in hazard impact.

Generalisation 3: The human response to hazards

- Examine the reasons why people respond to natural hazards in a variety of ways.
- Examine how people's responses are affected by the frequency and magnitude of natural hazard events.
- Education is often the most effective way of managing natural hazards. Discuss.
- To what extent do states of economic development influence the way people respond to natural hazards?
- Preventing a hazard event from becoming a disaster has more to do with human response than the magnitude of the event itself. Discuss.
- To what extent does the response to a hazard vary with the state of economic development?

- Perception of hazards is a key factor in the way people respond. Discuss.
- Examine the contribution that various groups of people make in the response to hazards.
- To what extent do responses to hazards vary with the state of economic development?

Generalisation 4: Global trends in hazards and hazard management

- Explain why some hazards are more predictable than others.
- The main aim of hazard management should be to reduce the effects of hazards, not to manage their causes. Discuss.
- Critically examine the view that natural hazards seem to be occurring with increasing frequency.
- To what extent can hazard prediction reduce the impact of natural hazards?
- The technological fix is the answer to the successful prediction of hazards. Discuss.
- Global warming is likely to have a significant impact in increasing the incidence and severity of impact for a range of natural hazards. Discuss.
- Examine the reasons for the decrease in deaths, yet an increase in damage, from natural hazards.
- Reducing vulnerability is key in coping with the increasing numbers of people affected by hazards. Discuss.

Sample essays

To what extent do responses to hazards vary with the state of economic development?

Total: 60 marks

Answer: borderline A-grade response

'A hazard is a perceived natural event which has the potential to threaten both life and property — a disaster is the realisation of this hazard.' The use of words such as 'perceived' and 'potential' by Whittow (1980) are indicative of the complex nature of hazards, of their unpredictability and of the destruction they may cause. However, Keller (1988) said that 'natural hazards are nothing more than natural processes. They become hazards when people live and work in areas in which these processes occur naturally'. This highlights the need for response and management systems to be put in place. Such systems and schemes depend upon a number of factors — in particular, the nature of the hazard, location, past experience, technological equipment, perceptions and political infrastructure. These criteria are linked closely to the economic development of the hazard area, whether it is global, national or regional. Different areas of the world have different responses to the occurrence of a hazard, but there is a dramatic contrast between MEDCs and LEDCs.

🖉 The definition is sound. However, the candidate should have discussed the LEDC/MEDC divide, rather than simply making a bald statement that there is a contrast in their responses to hazards. This introduction earns 6 marks out of a possible 10.

Costa Rica has a GNP per capita of $2160. There are 1179 people per doctor and 49.7% of the population live in urban areas. The location and climate of Costa Rica mean that it suffers from the threat of multiple hazards, such as floods, drought and landslides. Yet, even though there is constant danger, very little is done by either the government or local people to try and respond to hazards. People are fatalistic and rate employment and income as priorities over hazard insurance. This may be because 20% of the population live in poverty. This links in with the ideas of Marx — it is the poorest people who are affected most during times of drought or flood because they are the most vulnerable. A lack of capital means that there is little education about hazards, and hazard-management schemes are not implemented. The option of modifying the loss is a passive approach adopted by many LEDCs. Aid is one method of sharing the loss, but many governments are too proud to ask for help from organisations such as the Red Cross or the UN.

In contrast to the option of aid, the option of insurance is seen as the best way of sharing the loss. In California, less than 50% of the people have insurance greater than the mandatory state requirement. However, hazard insurance, particularly against earthquakes, is available. This is a sensible approach — 28 out of 30 of Lloyds of London's biggest insurance claims have been due to natural disasters.

California, in particular Los Angeles, is prone to earthquakes. In 1971, the Los Angeles dam situated in the San Fernando valley was hit by an earthquake of 6.7 on the Richter scale. Prediction techniques employed in the USA led to the evacuation of 80 000 people. Although it was believed that the dam would survive, one more aftershock would have destroyed the thin layer of dirt that lay between 80 000 people and 2 million tonnes of water. It was thanks to prediction methods and emergency response that so many people were saved. Another dam, costing $33 million, had to be built further downstream along the San Fernando River. It was tested in 1994 after the Northridge earthquake, which had far greater impacts on Los Angeles as bridges and highways were destroyed and fires due to broken gas mains raged for several days. The 'hazard-ready' approach in the USA prevented deaths and alleviated millions of dollars worth of damage.

Response, however, does vary within MEDCs. The magnitude of the event determines the response. Two of the richest and best-prepared nations have different responses to hazards. The Japanese believe that the hazard can somehow be alleviated and so invest in prediction methods and techniques to prevent the effects of the hazard. Although it applies some ideas of prevention, the USA focuses on the after-effects of the hazard and the immediate response, organising the emergency services and local people and using the National Guard to prevent looting and to maintain control.

Not all LEDCs fail to respond to hazards. The response to the huge eruption of Mt Pinatubo in the Philippines in 1991 was both well planned and well executed. A quarter of a million people were evacuated and it was originally thought that only 350 people were killed. The number of dead did increase as the remote Aeta people were brought in to evacuation centres. Their rural and isolated lifestyle resulted in susceptibility to disease and the death toll to rise to 800. However, the rapid response may be credited to the Americans. The Clark USAF base located on the island was involved in the prediction and subsequent evacuation because large numbers of American citizens live on the island. They have set up the Pinatubo Volcano Observatory to survey and monitor the activity of Mt Pinatubo.

Some MEDCs fail to respond to hazards. An avalanche struck Galtur, Austria in February 1999. After prolonged snowfall, 170000 tonnes of snow, moving at 80 mph, hit the village and 31 people were killed. This tragedy could surely have been averted if controlled response measures — such as moving snow with controlled explosions or evacuating the village — had been taken. This did not occur and resulted in embarrassment for local officials, who perhaps should have acted. This proves that it is not necessarily always money that determines the effect of a hazard.

Bangladesh is a hazard-prone area because 80% of the country is a floodplain and the government does not have the funds to build and maintain defences or control the threats. It has, however, improved its response to hazards. In 1970, a hurricane moved up the Bay of Bengal killing 300000 people. In 1985, a similar hurricane killed 40000, contaminated water supplies and increased salt levels in the soil, rendering agriculture impossible. In 1997, another hurricane hit the coastal regions, yet only 95 people were killed. This emphasises the fact that local people and the government had realised the threat of hazards and responded. With the help of foreign aid, early warning systems had been set up and people were educated about what to do to prevent hazards affecting them. Although excellent in principle, this sort of scheme does not always work.

In Mozambique, the NOAA has installed a similar monitoring and prediction system to that of Bangladesh. During the floods of 2000, 500 people were killed because, as a result of poor infrastructure and communications network, the message of rising water levels and increased rainfall did not get through to large numbers of the rural population.

On 26 December 2003, an earthquake of magnitude 6.7 hit the city of Bam in southeast Iran. Local authorities and officials had endured similar events before and could respond rapidly. However, the telephone connection to Tehran, 190 km northeast, failed and Bam was isolated. The high death toll (over 20000) was mainly due to poor communications and poor buildings. It has been stated that 75% of deaths caused by earthquakes are due to collapsing buildings. Increased rural–urban migration meant that the planners in Bam were finding it difficult to cater for their needs. Extra floors were added to original buildings and poor workmanship combined with poor-quality materials meant that a large number of buildings collapsed.

An earthquake of similar magnitude hit California on 23 December 2003. Only three people were killed and media representation was minimal. This is because of better building methods and materials and, more importantly, better building codes and enforced regulations. Walls in California must be 100% thicker than normal walls and joints have to be three times stronger. Buildings are designed not to amplify the resonant frequency of waves moving through the ground during an earthquake. Many buildings in the state have been retrofitted to be able to withstand earthquakes.

> *e* The candidate covers seven case studies involving earthquakes, volcanoes, avalanche and floods. The examples are taken from a range of locations and comparison is made between responses in MEDCs and LEDCs. There is evidence of precise, usually accurate research, which earns 12 of the available 15 marks. The level of understanding shown earns a further 12 marks.

Although there is a link between wealth and response, the unpredictability and complex nature of hazards make it very difficult to respond effectively and efficiently. Money helps, but it is only beneficial if invested in specific areas, particularly education and prediction, health care, industry and food production to help manage the consequences of a hazard. Having more money is likely to help a hazard response; in wealthier nations there is more to be lost and, therefore, more to replace. It is a fallacy that in LEDCs people have no money and no technology and so they wait to die. The response may be less hi-tech, but community preparedness and education can be successful, as the case studies show.

> *e* This conclusion is worth 6 marks out of 10. Although brief, it is meaningful and does refer back to the question. The essay is well structured, well organised and well written. This answer earns 8 of the 10 marks available for quality of written communication.

> *e* The total score is 44 marks, which just achieves a grade A.

Answer: borderline E-grade response

Different countries with different economic states respond to natural hazards in different ways. These responses are influenced by various things, but economic development is the one I will be focusing on.

Natural hazards include avalanches, drought, volcanoes and earthquakes.

> *e* This is a weak definition.

These hazards can have a short-term or long-term effect on people, destroying the economy, lives and buildings.

So to stop these hazards from destroying everything in their paths, we have to respond to them by trying to manage them.

Different countries manage hazards in different ways, but they all categorise the hazards into techtonic, meteorological and geomorphologic.

> *e* This part of the introduction is not well structured. There is no logical paragraph structure. The candidate has misspelt 'tectonic', which should not happen at A2.

Once the hazard has been categorised it then needs to be treated. There are three main points of management and response:

- modify the event — can it be managed by a technical fix?

- modify the vulnerability
- modify the loss — can people deal with the loss by themselves or can they share the losses?

   > *e* This seems like a pre-learnt list. The answer should be a formal essay, so should not include bullet points.

Once these points have been taken into account, countries can then go about responding and managing the natural hazard.

How a country manages a hazard depends on the state of economic development of a country — whether it is an MEDC or an LEDC.

   > *e* Despite the shortcomings, this introduction provides a sound framework for the topic. The candidate scores 7 marks out of the available 10.

For example, if we look at the Kobe earthquake, the people there managed the hazard much better than an LEDC like India or Iran. We know that to a certain extent we can predict earthquakes by looking at seismic action but we cannot actually prevent an earthquake from happening, so countries like Kobe, Iran and India have to manage the event as well as possible. With Kobe being an MEDC it has the money to be able to try and predict when and where an earthquake will happen.

The children of Kobe have earthquake drills at school every 4 years to try to make them respond better to earthquakes.

If an earthquake does occur then the people will be able to manage the damage and deal with it as quickly as possible.

   > *e* An A2 candidate should be aware that Kobe itself is not an MEDC — it is a city in an MEDC.

An LEDC (like Iran for example) is not so well equipped for an earthquake. The Iranian government and scientists are trying their best to work together to help manage these earthquakes. However, unlike Kobe with up-to-date technology and most buildings being aseismic, Iran cannot afford to make such changes. This shows that where earthquakes are concerned, economic development influences the way people respond.

Volcanoes are another natural hazard that people try to manage and respond to in different ways.

For example, Mt Pelee and Mt Pinatubo both erupted, but were managed differently.

The people of Mt Pelee thought something was wrong so they started to evacuate. However, at the time, government elections were taking place, so the troops were called in to drag the people back and keep them inside. Once Mt Pelee erupted, the troops were soon evacuating everyone again.

   > *e* This account is inaccurate as it mixes up the two case studies.

Whether this mistake was the government's fault or a lack of economic development we don't know, but one thing's for sure — if they had known that the volcano was going to erupt, they would have been able to evacuate sooner.

Mt Pinatubo, on the other hand, had special scientists and geologists watching the volcano to try to work out when exactly it would erupt. After working out that the volcano was going to erupt, they soon evacuated everyone. It was thought that educating the people of Pinatubo would be the best management strategy, so the scientists and geologists put together videos about the volcano for the people to watch and become educated about this volcano.

So whether the videos of Pinatubo actually worked or not, they still had the money to be able to get scientists in to assess the situation, whereas Pelee, with its lack of economic development or just poor judgement, did not manage the hazard very well at all.

Drought is the final natural hazard I will look at to determine whether economic development influences the way people respond to natural hazards.

If we look at the drought of Zimbabwe, the people there appealed for aid, which they got from South Africa and which helped them through the hard times of the drought. However, even after the drought, the people of Zimbabwe are trying to manage things in case this hazard happens again — they now have a water plan. They see what the water is being used for and manage it. Zimbabwe is an LEDC and was not able to manage the hazard without aid.

The UK, on the other hand, suffered a major drought but was able to manage the situation as it was ongoing. It introduced hosepipe bans and managed the little water that was left and kept sufficient amounts until the drought was over.

⌨ This answer shows limited research on three pairs of hazards. The accounts are descriptive and not always well selected (e.g. volcanoes). There is no attempt to state when any of the hazard events happened — not even the UK drought. The candidate scores 6 marks out of 15 for research and a further 6 marks for showing some understanding of the reasons for different types of management.

I think that by looking at how countries that are MEDCs manage and respond to hazard and how LEDCs respond to hazards, we can see a significant difference in management strategies. LEDCs just have to hope for aid and help, whereas MEDCs are able to manage the hazards well.

So I think overall that to some extent, economic development influences the way people respond to natural hazards.

⌨ This is a simple conclusion that is not linked back well to the chosen case studies. There is limited structure to the essay and the candidate has made some basic mistakes, scoring 4 out of 10. The style is basic and lacks sophistication. The score for quality of written communication is 5 marks out of 10.

⌨ **Overall, the candidate scores 28 marks out of 60, which is a borderline grade E. It is worth noting that there are only 16 marks between the first and second answers. The E-grade essay had a better framework to begin with, but the A-grade essay showed much greater knowledge of a wide range of case studies and there was good discussion of the reasons. It was also written in a much more sophisticated style.**

Summary

Features of high-quality essays include:

- an introduction that focuses on the issue, as opposed to providing routine definitions and long accounts about the choice of case studies
- research is up to date, wide-ranging and appropriately selected to focus on the chosen issue
- research that is enhanced by well-devised tables or statistics highlighting *comparative* features, such as, in this example, the parallel Bam and Californian earthquakes
- an understanding of conceptual diagram frameworks to which the case-study knowledge could be applied — for example, devising a response framework to look at before, during and after hazard events, to add focus to the answer
- conclusions that draw together research findings and essay arguments and then come to an evaluation based on the chosen title
- high quality of written communication, using a wide range of appropriate terminology, and a carefully crafted, well-structured exploration of the chosen title

Features of lower-quality essays include:

- an introduction that lacks both definitions and a clear focus on the question, although, in this E-grade example, efforts are made to list case studies
- basic research, lacking in precise detail, with no clear reasons given for selection
- some understanding of the question, but at a descriptive level
- a conclusion that is not well founded on the research
- a narrative style with poor use of terminology and a lack of sophistication